PROBLEM SOLVING,
THE SOLUTION TO THE PUZZLE

A constructive view that explains how to solve
problems at all three levels of a service or
manufacturing organization by using a simple
systematic approach.

MICHAEL RAY FINCHER

ISBN: 1451576056
ISBN-13: 9781451576054

Table of Contents

ACKNOWLEDGMENTS

I would like to thank my beautiful wife Melissa of 22 years for encouraging me to stay in the game and never giving up on me no matter what. For challenging me, teaching me, loving me, for picking me up each time I fell, and being by my side throughout this journey we call life. My two beautiful children, Kaila for helping me edit this book, Max for allowing me to be his hero, both for giving my life purpose. My parents Joe and Helen for giving me that launch pad and the tools I needed to fight for everything I have. My Brother Scott for being my inspiration and showing me that if I set my mind to it, I can achieve it. My Sister Terry for showing me that good is never good enough.

Thanks to Allen Etheridge for helping me edit and arrange the text to be more user friendly and for being an honest and true friend.

Thanks to you the reader of this project for allowing me to share some of my experiences and some of the tools I have learned along the way during my career.

Last but most, God for giving me the opportunity to experience all the situations I have ever been in that allowed me to share this information with you; and for putting this book in your hands so that it may change your prospective on problem solving and allow you to become more successful in your business or career.

Thank you all and God Bless!

INTRODUCTION

What is the definition of a problem? How can problems actually help us improve our environment? Is it necessary to hire highly paid highly skilled employees to solve our problems? The answers to all the above questions are just ahead. Problem solving is not always so complicated. Most of us would agree that a problem is actually a puzzle that needs to be solved. Problem solving is a matter of putting the pieces together so that the puzzle looks like what it is supposed to. Imagine if we were to put the same puzzle together time and time again. We would be very proficient at solving the puzzle the more we practiced. A problem can also be defined as someone or something not meeting the customers' expectations. We expected one thing and actually received something different. The problem is the difference between what we expected and what we actually received; or the gap between the two. For example, I expected my cell phone bill to be $75 and when I received it the bill was for $100. The problem is I now have to pay $25 more than I was expecting. I can solve this problem using the same approach I would if I had a problem at work or anywhere else for that matter. A majority of the problems we face every day in our environment can be solved by means of using very simple tools without using all the complicated statistical approaches we are lead to believe. This author has been in the quality profession for over 20 years and solved thousands of problems using the same simple approach. Of course there were times where more complicated tools were needed to solve the problem, but more times than not it was the simple approach that found resolution. There are numerous tools available for anyone that dare attempt to solve problems within their organization. Many tools are available such as six sigma, 8D, PDCA, and a plethora of statistical approaches that any organization may have at their disposal. The problem that most organizations have to deal with when attempting to enter into a problem solving task is they have no standardized approach established for problem solving. Most that struggle in problem solving end up trying to use a portion of many tools and fail to close the loop on the actual problem they are attempting to resolve. The reason the organization uses a portion of all the tools is mostly because that is what they have at their disposal within the available work groups. A typical work group of people will come from all over the industry with very different backgrounds. A typical work group would have been exposed to many different problems and many different problem solving tools throughout their career. An even bigger problem that exists is when these tools start to interact with each other and people become lost or confused. Some, if not all of the projects fail to follow the correct path and the actual problem is never solved because people eventually give up.

If we were to have a simple problem such as lack of parking at our business because additional employees were hired and we ask a team of three to solve it, we would get an array of responses from each team member. The team would no doubt use many different approaches to try and solve the problem. Maybe one member would focus on rebuilding the parking lot, another on how to reduce the number of cars, and the other on other alternatives such as off site parking. Any or all of these may be the solution but unless we direct the team to use a systematic approach it would only end up in an argument of who had the best answer. Problem solving is not an answer to a question; it is a solution to a problem. The building manager asked the team "What are we going to do about parking, we need more space?" What she is actually asking is how do we eliminate this problem and come to a resolution that will work for our current situation. Unless we totally understand what the actual problem is, offering solutions (or trying to answer the question) will only prolong the problem. If the team were to use a systematic approach to address the problem they may find out that solving it is actually very easy. Let's use a systematic approach and solve this problem. We must take it step by step. We will define the actual problem first and continue down the problem solving road.

Step 1: Define the problem (quantify the actual problem) – The current parking lot will only accommodate 15 cars but there are 20 employees that drive to work and need to park in the same lot. The business needs to be able to park 20 cars in the lot or provide an alternative solution. We have defined our boundaries and understand what we must do to solve this problem. Lot holds 15 cars, we need to park 20.

Step 2: Measure the problem (as it currently exist today) – The parking lot is 180 feet long and 32 feet wide and is marked with 15 parking spots. The spots are marked 12 feet wide and 19 feet long. The lot will currently accommodate 15 cars in this configuration if all cars park within the distinctive spaces. We have taken the actual quantitative data from our parking lot to see the current condition.

Step 3: Analyze the problem (research the data) – The city ordinance for a standard parking spot requires the width to be at least 9 feet wide and 19 feet long. The current parking spots are marked off at 12 feet wide and 19 feet long (3 feet wider than the city requirement). All 15 spots are equally marked at 12 by 19 feet. 12 feet wide multiplied by 15 spots equals 180 feet long (total length of our parking lot). We have analyzed the data against the known standard to understand the actual problem.

Step 4: Improve the problem (use the data to make a decision) – Using the city ordinance regulation, if we change the width from 12 feet down to 9 feet we would be able to accommodate 5 more cars in the same parking lot size. Current lot length is 180 feet long divide it by 9 feet wide parking spots and it equals 20 spots. The length of each spot of 19 feet is currently acceptable. We have solved our problem.

Step 5: Control the problem (ensure the improvements work) – The old lines on the parking lot were painted over and the new 9 feet aisles were marked off and fresh paint applied. The team had all employees park in the new designated spots to ensure enough room was made for 20 cars. All cars fit comfortably in the new 9 feet wide spots. To ensure the lines stay fresh, the team also added a line item on the building manages' to do list that requires the lines be touched up every March. We will now control our improvements by touching up the lines every year.

The following information (we will call a project) will demonstrate how to solve problems within the organization using a simple but systematic approach. It will show how those tools can be applied in a more global environment so that all levels of the organization can benefit. The information can be applied to the service industry, the manufacturing industry, and even beyond into ones personal life. Throughout this project we will focus on the 3 levels of an organization and how to apply the appropriate root cause analysis, and how to implement the appropriate corrective action at each level. This project will use many real world examples as to make the information as less complicated and as entertaining as possible. This project is a simple tool that can be used to teach all levels of management, engineers, and shop floor associates alike. Consequently, our main goal is to have an organization full of problem solvers and not just a few highly skilled highly paid individuals that we solely depend on to manage our processes and systems.

CHAPTER 1
SUPPLIER DEVELOPMENT,
A SYSTEMS APPROACH —

The word **supplier** in this context refers to any supplier internal or external to the organization. Supplier development can be applied to any supplier within the process flow upstream of the customer. Without developing the supplier and ensuring the upstream flow of goods or services are meeting their goals, we should not expect the downstream customer to be satisfied. The content of this text is not to define customers and suppliers, but to focus on the process approach tools that can be used to improve those that have been defined as a supplier within the total process scheme.

The systems approach for supplier development is a tool used to ensure the three levels of the organization are addressed when a problem has occurred. There are 3 levels to all organizations regardless if it is a service organization or a manufacturing organization. The three levels are:

1. The product – also known as the <u>output</u> (or the **y**), what the organization sells or produces.
2. The process – also known as the <u>method</u>, how the organization creates what they sell or produce.
3. The System – also known as the input (or the x), how the organization designs the process to create what they ultimately sell or produce.

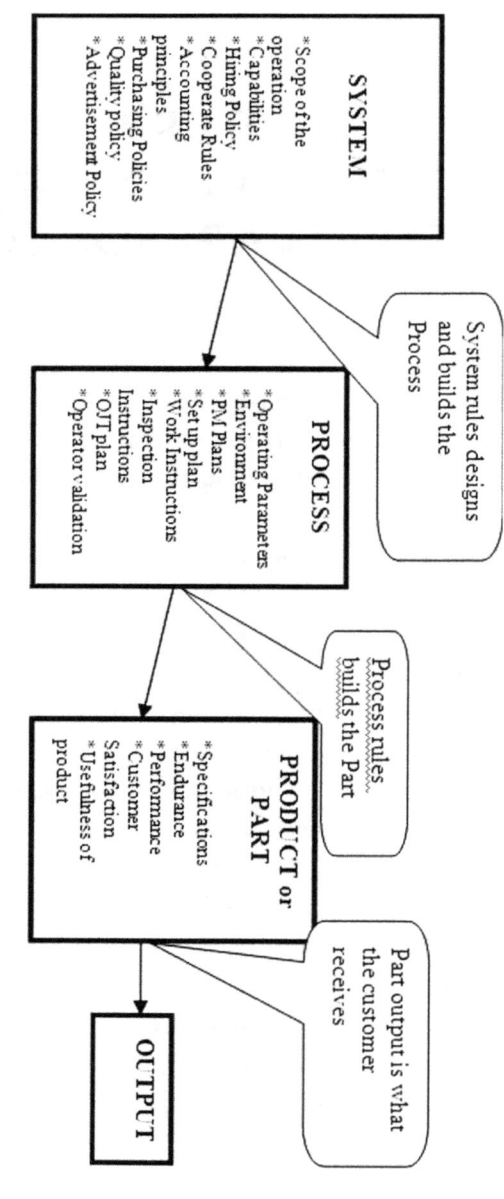

SYSTEM

* Scope of the operation
* Capabilities
* Hiring Policy
* Cooperate Rules
* Accounting principles
* Purchasing Policies
* Quality policy
* Advertisement Policy

System rules designs and builds the Process

PROCESS

* Operating Parameters
* Environment
* PM Plans
* Set up plan
* Work Instructions
* Inspection Instructions
* OJT plan
* Operator validation

Process rules builds the Part

PRODUCT or PART

* Specifications
* Endurance
* Performance
* Customer Satisfaction
* Usefulness of product

Part output is what the customer receives

OUTPUT

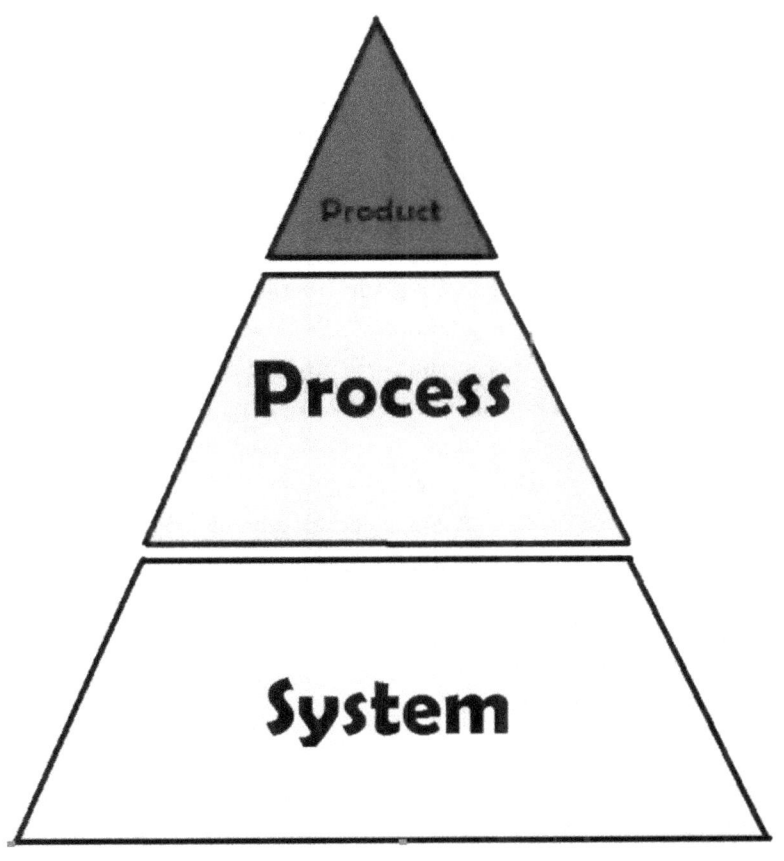

Each level of the organization will have different levels of problems and different levels of corrective actions. It is extremely important that all problem solving activities involve upper management. Each level of the organization requires different levels of managers' approvals to implement corrective actions and improvements. For example: Would we allow a shop floor operator to make a major change to the overall Quality Management System? Would we allow the Quality Manager to revise the production output to slow the line down so parts could be 100% inspected (verses sampling)? Would we allow a Nurse to revise the hospitals' way of administering medications? On the other hand, would we expect the CEO to revise the operator work instructions? I hope the answer to all the above is no. The organization should allow the appropriate levels within it to make the appropriate corrective actions at their level. That is not to say that everyone should not be involved in the corrective action process. If we allowed anyone at any level to make changes to every part of the organization we would have mass chaos.

Let's use this example: A Nurse decided administering medications to those patients that were quite capable of moving around on their own took way too much time, so she decided to revise the process. At the beginning of her shift for 30 minutes, she had the patients that could move about without assistance line up at her desk and serviced them like a production line. This saved almost 4 hours out of her day so she could now focus more time on those that needed more attention. Although this seems like a good idea, we can not know what the actual outcome of this process change will cause to the overall output. What if those patients that stood in line to get their meds fell during the process? What if the patients had different times they were supposed to take their meds? What if other factors were affected by revising this policy, like having to hire another Druggist to stage the meds in a method that would allow this process to work efficiently? What if a patient was sleeping during the distribution period and missed their meds? Obviously, the Nurse Manager would not allow such changes to the already defined process because the outcome could be detrimental to some patients. In the following text, examples will be shown of how the appropriate level of corrective action will be taken at the appropriate level within the organization when the product, the process, and the system fail to produce the desirable output.

PRODUCT (or part) level – typically the operator can implement corrective actions such as slight adjustments to the machine or reprioritizing patients based on their needs.

PROCESS level – typically engineers, process techs, or middle managers can implement corrective actions such as implementing additional quality gates, changing process parameters, or revising shop floor procedures.

SYSTEM level – typically upper managers, presidents, vice-presidents, CEO's, and plant managers can implement corrective actions such as changing hiring policies, changing the scope of the operation, or revising a purchasing policy.

SYSTEM TOOLS

Although this approach is not focused on the six sigma methodology exactly, there are basic principles applied for solving the problem and developing the supplier. The basic layout of the project that will be used is based on the DMAIC approach. DMAIC is: Define, Measure, Analyze, Improve, and Control.

Each of these will be explained and demonstrated in great detail in the following pages. Another tool that will commonly be used throughout this project is the 3P 5 why approach. The 3P is: Protect, Prevent, and Predict. This project will demonstrate how to apply all 3 P's to the three levels of the organization in which they apply.

PRODUCT level – **P**rotect

PROCESS level – **P**revent

SYSTEM level – **P**redict

CHAPTER 2
DMAIC (PRONOUNCED DUH-MAY-ICK) DEFINE —

Define is what the output is actually doing, or the results of all the upstream operations combined that is causing a defect or a problem. Maybe the organization is not experiencing what they consider problems, but would still like to improve; therefore they must define what targets need adjusting or improved before the task of correcting them can begin. Just like the example of the parking lot above, the business was in a situation where improvements had to be made when additional employees were added. The condition did not exist until additional employees were hired so an improvement to the condition had to be made. Define must be quantified; that is to say that in order to understand what the output is actually doing we must assign a quantifiable measurement to it. There must be some gage or some target that can be compared to the problem to see exactly how bad the situation actually is. A quantifiable measurement would be such targets as: Part specification requires the part to be 25.4mm +/- 1.0mm in length, Process tact time requires the process to run at 10 parts per minute, Plant Operating System policy requires the organization to maintain less than 1% lost time accidents per quarter. All three of these examples can be measured against an actual known standard, target, or written policy. It is not impossible to measure something without a target or spec, but it is impossible to determine how good or how bad the situation is without comparing the

7

measurement to the target or spec. If we were to state that our child's school teacher received a 5 on his assessment test, we do not know if this is good or bad. We would need to compare this rating against a target. If the target scale was worst to best 1 through 5 we know he was the best. If the target scale was worst to best 1 through 25 we would know that he was very bad. In order to define the problem in the define phase we must first know the target or go back to square one and set the target in order to continue on this project.

DEFINE THE PRODUCT –

If we were to define and quantify a *Part or Product* problem, we would state the following: The Specification for Part **ABC123** requires the length of the part to measure **25.4mm** +/- 1.0mm. On **1/1/2009** customer **ACME** reported finding **75 pieces** from lot number **12-15-08-SAM.** The actual measurements of the parts were **22.5mm**, which is **2.0mm** out of stated specification tolerance range. This statement allows us to investigate the part problem very quickly. We know when, where, and what the defect is according to the known standards. The only questions we can not answer at this level is why the 75 parts were out of specification and the source of the failure. We have however defined this particular part problem and can hopefully eliminate it very quickly, or at least find the root cause of this issue much quicker than if we merely used qualitative data to define it.

Qualitative data on the other hand is not precise (not discrete) and should not be used to define a problem unless absolutely necessary. From the example above if we only used qualitative data, the problem statement might read like this: ACME received numerous brackets in Jan, 2009 and reported that some of them were short. By only using the qualitative data we still know that the customer received bad parts but we do not know part number, lot number, quantity bad, how far the parts were out of spec, or when they received the problem.

How difficult would it be to solve this problem using only qualitative data? Could it be solved accurately and in a timely manner? By using this data our problem solving team would scatter in many directions just like the team did in the example above for the parking lot issue. We need to narrow down the problem, or define it so that we use our team as efficiently and accurately as possible to resolve the problem. We as champion problem solvers need to provide the rest of the team with the road map and directions on how to arrive at the destination or conclusion of the problem. If we can define the problem by providing the rules, the team can easily compare the actual data results to those rules to see how bad the problem actually is and what direction they need to go to solve the problem.

DEFINE THE PROCESS —

Let's continue to investigate and define the problem for a *Process* failure using quantitative data. An example of a quantifiable process problem might read like this: 75 pieces of Part ABC123 were 2.0mm out of spec (we know this from the Part level) because the **left front locating pin** in tool **XYZ789** broke during processing on **12-15-08**. The pin broke at piece number **724** out of an 800 piece production order. Again, from this quantifiable statement we know, when the pin broke, where it broke, and what exact pin broke that caused the defect. The only question we can not answer at this level is why the pin broke that caused the defect. Do we know what the process rule was for this condition? Obviously it was that the pin must not break during processing, right? How do we know that? If we had a process rule that stated the pin must not break during processing then there must have been another failure. Let's say that the team investigated all the process rules and did not find any target or requirement that required the pin to be checked or replaced before it broke. How can we compare the actual condition to a rule that does not exist? We can not. In this case it is obvious that we have not defined a target that has a detrimental impact on the part quality. We have no rule established for the actual pin that broke; therefore we can not define the actual condition against that rule. Before we can solve the process problem, we must define the rule for this condition. We must make it a rule that the pin must be replaced before it breaks during processing. At what interval should we change the pin out to ensure it does not break during processing? The answer lies ahead in the Measure phase of this project. At this point, we have defined that there is no process rule that requires us to change out the pin before it causes defects. Therefore the process will continue to make defects as more pins break during processing until we can establish a rule to prevent this defect.

DEFINE THE SYSTEM —

To continue on to the *System* level, we can further define and quantify the above issue using quantifiable data. An example of the systems problem might read like this: On 12-15-08 tool XYZ789 left front locating pin broke during an 800 piece production run (we know this from the Process level). After investigating the issue, it was determined that the pin had been classified as an expendable part by the Maintenance Department during production launch. The endurance data had been collected for this pin and determined to have a change interval at **5000** cycles. The data was not used to update the preventive maintenance plan resulting in no rule to require pin changes. Now that we are finally at the systems level of actually defining the problem, we can start to solve the problem using the systematic approach.

Customer received 75 pieces out of specification.

PRODUCT LEVEL – Parts were 2.0mm out of tolerance.

PROCESS LEVEL – Pin broke causing the defect to occur.

SYSTEMS LEVEL - Preventive Maintenance had not been established.

As you can see from the simple exercise above, without defining the problem for all three levels of the organization (all the way to the systems level) we can not accurately asses the problem; therefore we will not be able to implement the controls to resolve the problem. If we simply change the pin out we have only eliminated the *part* problem. If we merely update the preventive maintenance plan we have only eliminated the specific *process* problem. BUT, if we revise the system to develop a formal operating standard across the organization we will have eliminated the *systems* problem all together.

It would be close to impossible to determine what the systems problem was in the exercise above if we failed to first define and quantify the part problem and then the process problem. By quantifying the part problem first, we can then move to the process, and then to the system using this simple road map of problem solving.

As described above, each level of the organization will have to follow the same systematic road map to solving the problem all together so that it will not reoccur. On the same note, we were not able to see that we had a systems problem until we defined the problem on the part level first.

Many times when a corrective action team gathers to help solve a problem, someone on the team nearly always jumps to the conclusion without properly defining the actual problem. They are merely providing an answer to the question and not a solution to the problem. This type of activity needs to be avoided or the team will scatter in all directions as they did in the parking lot example. As the team champion we must challenge those answers and push for solutions. The Champion of the team should ask one simple question to those with only answers to the question. "What tool did you use to define your conclusion?" How could these persons know what the actual solution to the problem was at the Part level, Process level, and Systems level unless they understand the actual targets we were aiming for?

In the parking lot example let's define the targets for each level:

The input (x) = 5 additional employees were hired.

The output (y) = 5 employees had no parking space at the business.

Product Level – 20 employees, only 15 parking spots, off target by 5 parking spots.

Process Level – 5 Additional people were hired and the parking lot would not accommodate an additional 5 cars.

Systems Level –Adding more parking spots before additional employees are hired is not part of the current hiring policy within the organization.

THE UNKNOWN –

Some issues like the problem described above are fairly easy to define while other more complicated problems are very difficult to define. How do we define those problems that are more difficult? The answer is simple; we use exactly the same steps as described above. Every problem will have a result or an output (x). The output is known as the product.

Every system will have an output. Remember, the system is what builds the process, the process is what builds the part, and the part is the ultimate output or product. So, using the same roadmap we will always start with the product and work our way upstream to find where the problem exist in the process as well as the system. We should never stop until we have reached the system level problem because this is where we prevent problems on other like processes (preventative actions). If we had stopped at the process level in the parking lot example and just added additional parking spots, what happens when the business hires more employees? We will continue to solve the same problem over and over unless we address the upstream system problem and implement the appropriate solution.

Let's look at another example of a more difficult problem. The President of the company reported to us in our daily review meeting that the new part we launched 2 months ago is now 3 days behind schedule on outstanding orders and we will have to work the next 2 weekends to make up the late orders. The problem is that there is not enough money in the budget to pay all the overtime, but unless we make up the orders we will lose the account. Our corrective action team was assigned to solve this problem so that we make the orders and save the business. Do we have enough data at this point to deploy our team to the appropriate levels of the organization to solve this problem? Has this problem been quantified at each level? The answer is obviously no. So, where do we start? Again, easy answer, we start at the DEFINE phase using our systematic process approach. We must first accurately define the problem at the product level using quantifiable data. The team should be asking questions to narrow down the problem such as:

What is the quoted volume on this part (how many parts did we agree to make each week)?

Exactly how many pieces per hour, per day, per week are we currently producing against the quoted rate? What is the current scrap rate target and current scrap rate output?

What is the quoted down time target and current down time output?

Based on the original process capability study, where do we compare to current outputs?

Notice that all the questions above focus on the product targets (targets should always be measureable). We are trying to compare the current output (condition) to the set standards so we can narrow down our scope. Let's say the team came back with these answers:

What is the quoted volume on this part (how many parts did we agree to make each week)? ***40,000 pcs per week.**

Exactly how many pieces per hour, per day, per week are we currently producing against the quoted rate? ***800 per hour, 6,400 per day, and 32,000 per week (negative 8,000 per week we fall behind).**

What is the quoted scrap rate target? ***less than 10 parts per day**

What is the current scrap rate output? ***225 pcs per day (215 over target)**

What is the quoted down time target? ***30 min per day**

How much down time have we had compared to the quoted target? ***30 min per day (on target)**

Now we have a path to follow. The current scrap rate is not reaching the target so obviously we will peruse that issue. To further narrow down the problem we find out that the reason the scrap rate is so high is because the parts are out of print specification by 2.0mm. Once we are here, we move on to the process level and there we discover that a pin keeps breaking causing the parts to be out of spec. From there we move to the system and discover that the PM program has not been defined to require the pin to be replaced on a prescribed basis (based on the capability data of the pin). As you can clearly see, one small part problem can result in a catastrophic system failure and effect the entire organization.

PROBLEM STATEMENT (defining the problem, quantified): - We are currently 12,800 pcs behind schedule on part number ABC123. Our quoted scrap rate goal is 5 pcs per day but the current scrap rate output is 225 pcs per day. 225 pcs per day are scrapped for being 2.0mm out of print tolerance. The out of spec condition is due to a pin breaking in the tool. We have defined our problem as a dimensional issue with the parts therefore we are not producing enough parts to satisfy our open orders with the customer.

PRODUCT LEVEL – Parts were 2.0mm out of tolerance.

PROCESS LEVEL – Pin broke causing the defect to occur.

SYSTEMS LEVEL – Preventive Maintenance had not been established.

By taking time to clearly define the problem we can react quicker and more accurately than trying to solve the issue using only qualitative data. The Define phase is the most important part of the entire problem solving phases because it is the starting point. If we get off to a bad start and follow the wrong path it could take weeks, months, or even years to solve the actual problem that is having a negative impact on the product / output.

CHAPTER 3
MEASURE –

The Measure phase can be rather complicated if we fail to follow a standardized approach just as we did in the Define phase. Once we enter into the Measure phase we should have already determined what we will be measuring from the Define phase. In the Define phase we collected and gathered all the specifications, parameters, and target goals. In this phase we will asses the measurement system, take measurements of the items, and we will also assign the accountabilities to each measurable. In other words, if there is a target that needs to be measured, we must first have faith in the gage or measurement system, and then assign a specific person or department to take the actual measurements of the current condition.

Another common finding in the Measure phase is that we find a target has never been defined. For example: The customer calls to complain that the parts they received all have a small imperfection on the top. From the Define phase we know that we must quickly quantify the problem, so we pull the part specification out to review it. While reviewing and studying the specification, we find no requirement on the part appearance that we could apply to this condition. How can we measure and quantify this defect if we have no standard or requirement to compare the part to? Another example: A Nurse is called to the Managers office because a patient had complained that his room temp was too hot. The Nurse Manager instructs the Nurse to adjust the room temp accordingly.

After retuning to the floor and searching for several hours the Nurse can not find the required setting parameter for the room temp. What should he do? How would he be able to bring the process back into control if he does not know the temp parameters or the rules? How would we handle that situation in a real world condition? If our customer is complaining about an issue that has not been defined, how can we properly measure that defect and state that it is in fact a problem? Think briefly about these questions. Who is ultimately responsible for developing a specification? Who makes the rules? Who will benefit most form the specifications? Now, stop and think about the above questions for a minute.

Let me rephrase the questions a little to make it more obvious. If you are shopping for a hammer in a retail store you will find many different selections at many different prices with many different options. You are in the market to buy a hammer to hang a few pictures around the house and plan to use it 1 or 2 times per year. Would you spend $285 to purchase the contractor grade hammer that is built for long term abuse and endurance? You would probably be more likely to spend $5 on the cheaper grade that will perform exactly the job you need it to perform. Let's say you purchased the $5 hammer and took it home and ended up building a deck with it. After hundreds of nails being driven with it and week's worth of abuse, the hammer eventually breaks. Will you return the hammer to the store for a refund? You are probably more likely to throw the hammer in the trash and replace it with a new one. Let's say you decided to purchase the $285 contractor grade hammer and the same situation occurred. Would you return the hammer to the store for a refund? You are probably more likely to return the $285 hammer because it did not meet your expectations for the price you paid for it. The term "You get what you pay for" is true in most cases. The price you pay for the hammer is a true indication of what the customer should expect in return for the investment. So, let's go back to the original question of what we should do if there is no specification established. Let me ask the question again. Who is ultimately responsible for developing a specification? I hope you have thought long enough to answer that, the consumer or customer does. So, if the customer complains that the imperfection is not what they expected, we should resolve the problem so that the customers' needs are met. If the temperature is too hot in a patient's room, we would simply ask them what temperature they would like it set to. If we do not resolve the problem, the customer may decide to take his business elsewhere and we would ultimately lose anyway.

In the example above, we would simply get with the customer and quantify their expectations on the imperfection and then make it part of our specifications or rules. In other words, we would step back one step to the Define phase and define the actual target. We would then get a quantifiable measurement on how big the imperfection could be (if allowed at all). Once this has been decided by customer and supplier we then make it the rule (we define the target). If we have any further imperfections on the part we will now be able to measure them using the quantifiable specification to compare spec to defect because we have now defined it. Again, we are not able to solve the problem unless we use the appropriate standardized steps to go from problem to resolution. As demonstrated, it is not possible to quantify how bad the defect actually is unless we first define the target. Once we define the target we can then see how bad the problem actually is. **NOTE:** Let's not focus our resources on why the spec did not exist. The customer may not always know every possible defect that could occur on the part they purchase. The customer is not the expert on the part, the supplier is. If the customer starts receiving parts that have defects that they reject, it is for a reason. The reason is the consumer will not be satisfied with the defect and may not purchase the end product. We are all consumers and demand a certain quality and design, although we may not always completely understand the design.

We will study the 3P 5 why approach later in this project in the Analyze phase. It is however important to understand the basic principle of the 3P's before going further into the measure phase. It is important to note that the 3P's directly relate to the 3 levels of the organization on a broader level. The following definitions of the 3P's will help the reader understand in more detail which P belongs to which level of the organization. Remember, the 3 P's are: Protect, Prevent, & Predict.

LEVELS OF THE ORGANIZATION –

PRODUCT level – **Protect**: product characteristics, the actual measurement of the product as compared to the defined specifications. Protect the customer from the defect by measuring and detecting the part failures.

- Examples of **Product characteristics** are:
    - Dimensional
    - Appearance
    - Performance / endurance
    - Under/over torque
    - Material hardness

PROCESS level – **Prevent**: process characteristics, the actual measurement of the process condition as compared to the defined operating parameters. Prevent the problem from occurring in the process; prevent the process from making a bad product.

- Examples of **Process characteristics** are:
  - Preventative Maintenance intervals
  - Equipment condition
  - Speed of equipment
  - Cure time of material
  - Mold Temperature

SYSTEM level – **Predict**: management characteristics, the actual measurement of an operating system as compared to the defined policies for the organization. Predict the potential problems by implementing system controls to collect and analyze the data before a problem occurs. If the problem can be predicted from actual data, the process can be designed to prevent the problem from occurring.

- Examples of **System characteristics** are:
  - Control of Non-conforming material
  - Reaction to out of control processes
  - New Product launch system
  - Training and Hiring policies
  - Accounting practices

SETTING THE TARGETS –

This project is not a precise detailed instruction on how to set targets within the organization. All aspects of the organization should be considered when targets are established. Remember that not all targets within an organization are as important to some as they are to others. A Quality Manager may not be so concerned with the overall lost time accident goal while the HR Manager may not be concerned with the overall quality performance goal. Both goals do interact with each other at the systems level whether the two managers can see it from their position or not. If we do not meet our lost time accident goal we may face severe penalties from our insurance provider resulting in a higher premium (just to mention one output). If we do not hit our quality performance goal we could lose business from the customer or have high scrap rates. Both of these goals have an output target of <u>profitability</u> in common. One target is just as important to the organization as the other. The two targets must also be aligned with each other so that one does not directly have impact on the other. If we do not have operators to run the line because 10% of them got injured and are out of work, we would probably have to hire temporary untrained operators to fill those positions. Untrained operators could potentially cause quality problems and affect the quality performance goal. So, it is always important to set the goals based on the upstream downstream methodology so that everyone has the opportunity to accomplish their mission within the established target. The basic concept for setting targets is to ensure that no target is set anywhere upstream in the organization that would cause another downstream operation to fail. The same basic principles apply when setting targets within the process.

TYPES OF PARAMETERS –

There are 2 types of parameters that you need to be aware of when attempting to measure; controllable parameters and noise. Controllable parameters are those parameters that have controls established such as feed, speed, time, temperature of a process. Noise is a parameter that currently has no controls such as outside humidity surrounding the process, UV rays, or temperature. Both parameters can be measured, however controllable parameters are the only parameters that can be adjusted. It may be possible to turn noise into a controllable parameter and then be able to adjust it. For example, building an enclosure over a process and installing a central unit to control the temperature environment. If temperature had been a factor in a problem, the user can now remove that factor by adjusting the temperature to the desired level and eliminating the problem (or noise as a factor).

MEASUREMENT TOOLS –

Depending on the situation, we may select a variety of measurement tools to measure with. The first practice of the measurement phase is to first ensure the measurement system is accurate and free from bias. Bias is described as an unfair decision. If for example you were to ask a Quality Inspector to measure a precision part with a tolerance of +/- .05mm using a tape measure, we would expect to see a bias in the results. It would be unfair or unrealistic to presume that we would get an accurate reading using a tape measure when we should have been using a micrometer. Same is true when measuring systems. If a Production Managers' pay was based on first time yield, would we allow him to collect and report the data any way he wished? Of course we would not put the company or him in that situation. We would define exactly what and how to measure the first time yield to prevent measurement bias.

By ensuring the measurement system is accurate we will have confidence in the results therefore allowing us to make the appropriate analysis of the data. Every gage that is used to measure, whether it is a gage used to measure a part, a process, or a system, can be calibrated and verified for accuracy. A typical gage for measuring part features can be calibrated using known standards. A gage R&R can also be performed on those gages rather quickly to ensure accuracy across users. A qualified

Quality Engineer should be able to perform a gage R&R with little to no problems to verify almost any gage used to measure part features.

Some of the most difficult gages to calibrate are those that are used to measure a process or a system. Let's take for example a gage that would be used to measure the turnover rate at a particular company to determine if the company is a good place to work. If our goal was less than 1% turn over rate within each quarter, what type of gage would we use to measure that target? We could compare the number of employees that were employed within the Qtr and compare it to the number of employees that left the company during the same Qtr. Does that sound like an accurate measurement? Would it include the number of employees that retired? Would the measurement system tell us if the people that left were long time employees or employees that had only been with the company a short time? Would it tell us if the people that left were fired, or quit on their own? Would it tell us if the employees were replaced within the same Qtr, therefore would not be counted in the overall percentage difference? As you can see, we could measure the target in many different aspects that would tell us many different things. The key point to setting up a measurement system is to determine exactly what we want to measure, what is the output we are looking for, and then apply the appropriate gage to collect that data to determine if we actually hit the target or not.

First, we must decide what measurement system to use before we attempt to calibrate or verify its' accuracy. How can we make certain that the measurement system we have selected is accurate? This can be done in a number of ways, but the best practice is to use multiple levels to measure the same target. In a process we measure the width of the raw material before we load it into the press and stamp a part out of it. After the part is formed, we will measure the width again using a different gage at a different operation step. If the width is critical, we will probably measure it several times throughout the process operation steps using a number of measurement tools to verify the width is correct.

Most organizations use the same method to measure their Quality Management System (QMS). They will set up measurement tools in a downstream configuration and measure the same target numerous times to ensure it is accurate. To look at the entire measurement system all at one time, the organization will conduct a management review meeting. A management review board will meet on a regular term and basically compare notes. An entire measurement system can be confirmed in this manner by looking downstream at the actual output or results of what each are doing. In the example above, we could actually confirm our turn over rate was meeting the goal by other measuring systems that were in place downstream. If the Production Manager reported during the management review board meeting that he failed to hit a target and was unable to run one of his machines last week because he did not have the employees to run it; we would quickly see that the turnover rate did not hit the target. How do we know the turnover rate did not hit the target? Obviously, we would not have set the turn over rate target to 1% if we knew 1% would result in a failure of the production target because there would not be enough employees to run all the machines. So, if the HR Manager came to the meeting and reported that the turnover rate was only ½%, we would quickly conclude one of two things; One, he was modifying the data so he hit the target and met his goal, or two, the measurement method he is using is not accurate and therefore should not be trusted. Just as we do on our process, those outputs that are critical to the product or the overall system we implement multiple layers of measurement systems to monitor the actual results.

Now we understand that the gage must be accurate in order to measure the output, the next step is to actually measure the output. <u>We must measure the output in its' current condition before any changes are made</u>. If we change the system, process, or product prior to measuring it we will introduce more variables into the collection that we may not be able to segregate during the analysis phase of this project. The main purpose of measuring is to try and isolate the variable that is causing the problem and determine exactly how bad the problem actually is. If we can find the variable that is not performing within target we can easily eliminate or improve it to remove the problem.

Remember that we will always start at the product level when embarking on a problem solving mission, work our way up through the process, and then to the system. That means we will need to take measurements at all three levels in order to completely solve the problem. Below are just a few examples of some of the most common measuring tools used at each level.

PART LEVEL Measurement tools (<u>P</u>rotect):

Fixtures – used for quick in-process inspections

Statistical Process Control (SPC)

Hand held measuring devices such as calipers, micrometers

Endurance results / performance data

Warranty data / Customer data

Surveys or product audits

PROCESS LEVEL Measurement tools (<u>P</u>revent):

Cycle Times

Downtime data

Scrap Data

Process audit results

Rework data

Machine Parameter output data

Environment condition

SYSTEM LEVEL Measurement tools (Predict):

Profit level

Employee turn over rate

Lost time accidents

Delivery of product performance

Quality Performance

Cost of materials

Employee involvement

MEASURING THE PRODUCT LEVEL (PROTECT) –

The task of measuring the product is one of the easiest levels of the measuring phase. This level is as simple as pulling the product specifications and comparing the actual product to the defined specifications. Remember from the Define phase that if we have yet to define those specifications we must move back one step and define the actual specifications before we continue to the Measure phase.

A very valuable tool to use in the product level measurement phase is the use of a Product or Process Map. The map should list each step in the process that requires a measurement point, how often to measure, what tool is used to measure with, and what actions to take if the product is out of specification. *Please refer to Chapter 8 for more details on Process Mapping.* The Process Map contains both Product Characteristics and Process Characteristics indicating the document is primarily focused on the product and process levels verses the system.

Use the map as a checklist and measure each product characteristic that is specified within the process. The best practice to use when measuring the product is to perform a full dimensional layout. The reason this is suggested is because there may be more than one characteristic that is causing the failure. For example, a bracket width may be 2.0mm out of spec in the finished part stage so the team may elect to look at the locating stop pin that controls the width. The actual problem was the leg bend angle that has a 90 degree bend specification measured at 80 degree. Because the bend was 10 degrees out of spec it caused the overall width to be 2.0mm short. If we would have stopped our measuring efforts we would have gone down the wrong path to solve this problem, and possibly created more problems.

Unlike the Process level, the Part level is strictly assigned to the actual product (or the output of the process). There are fewer characteristics to measure in the Product level than in the Process level. This is because the process my produce many different outputs or products from the same equipment, but the product is the actual finished unit and is always unique. The process that produces pencils could potentially make numerous colors from the same equipment. If we were to have an issue with only the blue pencils we would focus our measuring task to the blue pencils only. After we measure the characteristics of the part we would then move to the actual process characteristics, and then on to the systems characteristics. Using this standardized approach we can quickly define the part issue by quantifying the problem at the part level first, and then move on to the process level. The reason we measure the part level first is so we can implement controls to protect the customer. Typically the organization can not afford to suspend production until the process and system problems are solved. By measuring the part first, we can determine a temporary counter measure such as containment or rework.

More than likely, your customer will provide you with the actual part failure mode. That is to say that most problem solving activities will not be launched until the customer complains about the product condition. The data that will be received from the customer will more than likely be qualitative data and will not provide enough information to understand exactly what the problem is or where it is originating from. The data from the actual product should be verified so that solid quantitative data can be collected. The basic task of the Product level measure phase is to take the information that came from the customer and quantify it so the data can be analyzed in the next phase of this project. Although the product measure phase may be the most simple, it is the most important because the output is only what the customer sees. Typical customers will never see the actual system or the actual process that is responsible for producing the product. They will only see the output of those two levels in the form of the finished product.

MEASURING THE PROCESS LEVEL (PREVENT) —

There will be many different places that need measuring during a problem solving or continuous improvement project. One of the most valuable tools that should be used during the measure phase of the process level will be the process map. A good process map should list each operation within the process step-by-step from start to finish and the parameters for each step. *See Chapter 8 for more details on how to create and use a process map.* Just as with the Product level measurement phase, the Process level is focused on one particular characteristic; the Process Characteristic. Using the process map, the team should actually walk the entire process using it as a guide and making note of any discrepancies they find. The team should pay particular attention to any hidden factories. Hidden factories are one of the most common places to find problems and vast amounts of variation. A hidden factory is an operation within the process that is being performed and is not part of the normal process design. For example: The process map lists the steps of the process as such:

Operation 1 – Mold part

Operation 2 – Trim Part on all edges

Operation 3 – Place part on cooling fixture 5 min

Operation 4 – Inspect part

Operation 5 – Package part

After the team has performed the process audit, they discover the actual operations are as follows:

Operation 1 – Mold part

Operation 2 – Trim Part on edge 1 and 2

Operation 3 – Place part on cooling fixture 5 min

Operation 4 – Trim Part on edge 3 and 4

Operation 5 – Inspect part

Operation 6 – Package part

Operation 2 and Operation 4 are not part of the approved process design, therefore could be causing defects being entered into the process output. The two operations were added by the operator because they found it easier to trim the part differently from the approved process design. Because these two steps are different from the process design we would conclude that there is no capability data to prove that the revision actually results the same or better output of the process. Even though these steps are not listed on the process map, we still need to measure the output. Remember to measure based on the current condition. <u>Do not correct the operator then measure the output or you could</u>

miss an opportunity to find the problem that the customer was complaining about. Many opportunities are missed because the process was adjusted during a process audit and the problem was never measured in the current condition. If we do not address the issue and put in controls to prevent it in the future, the issue will reoccur time and time again. We could potentially have to go back and solve this problem numerous times if we continue to correct the process before we measure it.

One of the best questions to keep repeating during a process measurement activity is: Where in the process has the potential to produce the most amount of variation that has the least amount of controls? If you were only looking at the process map and you noticed there were a total of 10 operations listed, 9 of which are controlled by computer and 1 that was operator dependent; which operations would you be more apt to look for the problem? Which operations out of the 10 could you quickly asses to have the least amount of controls but could potentially have the most amount of variation? Of course it would be the operation that is operator dependent.

A good note to remember while measuring the process is to look for those operations that allow for mistake and are not full proof. If we design a process and allow the operator the INTENT, OPPORTUNITY, and CAPABILITY of making a mistake, eventually they will.

- **Intent** – objective, unknowingly make a mistake:
  - "That was the way I understood it to be done"
- **Opportunity** – chance, favorable circumstance that allows mistake:
  - "I had many choices, so I picked the best one I could"
- **Capability** – means, actual ability to carry out the mistake:
  - "My equipment can be adjusted, so I adjusted it to finish quicker"

Humans often times get distracted or become lackadaisical in the tasks they are performing and skip steps, add steps, speed up, or slow down. If we set up a process to make chocolate flavored milk and we have dispensers with water, juice, cola, and milk available to the operator, what is the probability that the operator will eventually pick up the wrong dispensing hose and fill the container with cola or something other than milk? We would have to do the actual math on this but I can say even one time would not be acceptable to the end customer. Can you imagine getting a mouth full of chocolate flavored orange juice?

MEASURING THE SYSTEM LEVEL (PREDICT) –

The final task in the measure phase is measuring the system. Measuring the system is by far the most complicated out of the three levels. Our focus for measuring the System level is to find those system targets that have a direct impact on the final product and measure accordingly. There are typically two main failures at the system level that we can measure. The first most typical problem is the target has never been set. As described in the Define phase, if the target has never been set there will be no one assigned to be accountable for controlling it. The second problem is the target is not being analyzed appropriately. Obviously, if the target has not been defined the target will not be measured or analyzed. Sometimes the organization will set the target but will fail to analyze the actual data. The organization will not be able to find any problems at the system level unless the data is analyzed. For example, the organization will set a target of less than 1% turn over rate per quarter. That sounds like a very good target to include in the overall operating system to present to the employees. This target shows the employees that the organization cares about employee longevity and has set a target to ensure employees are happy and continue to work for the organization. The target being defined does not necessarily mean that it will be analyzed or controlled. The actual employee turn over rate could be as high as 10% before anyone would react. Unless the target is directly related to another downstream goal it may never be considered until the down stream target (which is usually considered more important) misses the goal. From the previous example, the Production Manager may miss a production goal because he had no employees to run the machines. He found out later the reason he missed his target was because the turnover rate missed its' target.

We would typically never find that the systems level target missed the goal unless the final product was directly affected. The reason is because we can not always see the actual output of the system until the final product is produced. That is to say we typically do not react to the system level until the customer complains. It is a common practice to think we make our profit on the actual finished product therefore the product level is the most important. This is a very bad misconception by managers. If the system can build a process that prevents defects, defects would never reach the customer. No defect means a greater profit margin. It is very common for an organization to attempt to inspect defects out of the process rather than design the system and process to prevent the defects in the first place.

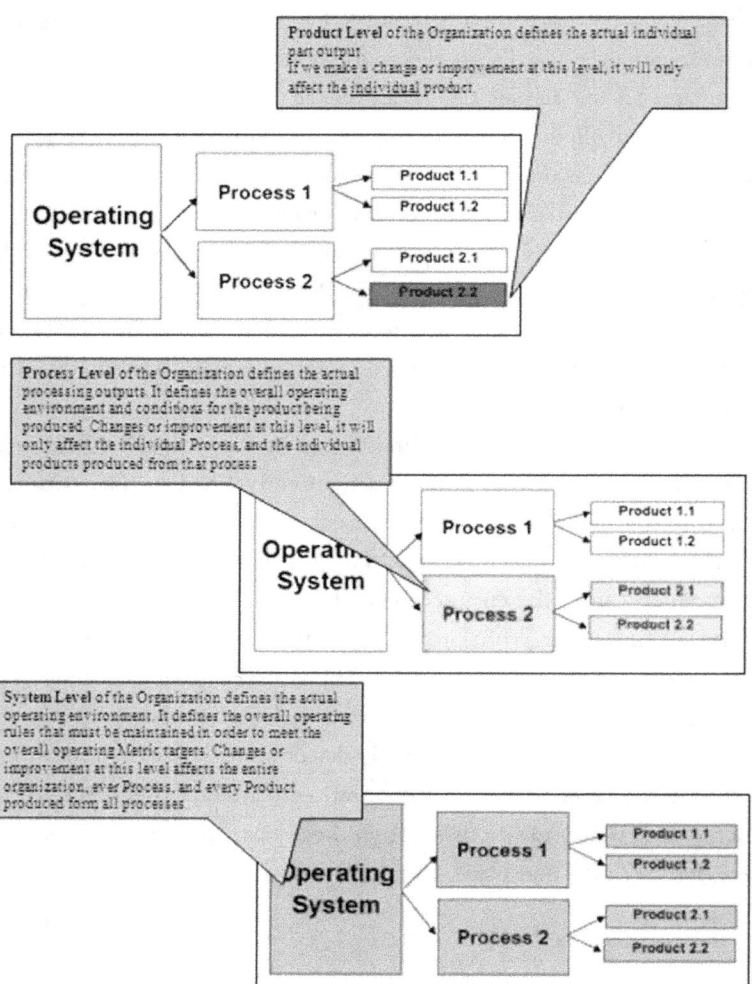

The simple answer as to what should be measured at the systems level are those systems that have been identified downstream of having a direct impact on the final product. Through proper investigation of down stream operations, the trail should always lead us upstream to the system that actually caused the defect; or should we say allowed the defect to be produced. We should always ensure our measuring tools are calibrated before taking the measurement, the measurements are taken at the actual condition before improvements, and the data is collected from all three levels of the organization. If we fail to measure all three levels of the organization we will abandon an opportunity to improve the overall environment of the organization setting ourselves up for many more reoccurring problems in the future.

CHAPTER 4
ANALIZE —

At this point in the project we have defined the targets that need to be measured, we have taken measurements of those targets at the product level, process level, and the systems level. In the Analyze phase we will use the collected quantitative data to ascertain from it our next steps. The most important part of the analyze phase is to determine what the actual problem is on all three levels (individually) so we can find the source of the problem. Data analysis is a very important part of the entire project because at this point we will decide what path we need to follow to implement the appropriate correct actions. If we fail to see what the data is actually telling us, we could fail to solve the problem or implement the wrong corrective actions. At all three levels of the organization, there are numerous tools available to analyze data with.

As with the Measure phase, the product level will be the most simple of the three. The product level is as simple as comparing the collected data to the product specifications. Although some applications may be more difficult to measure and analyze than others, the basic principles and basic tools will almost always apply. Again, this project was not designed to educate the reader on all the possible statistical analysis tools available to measure and analyze product characteristics. The process analysis will become slightly more complicated than the product analysis because it will be dealing with many more complicated measurements; such as mixtures, formulas, pressures, feed, and speeds. Although more complicated, the same basic principles still apply. The system becomes even more difficult to analyze because there are many points that will be collected that are considered not to be as important as the actual results of the product. Remember from the Define phase that it will be very difficult to analyze data if the target was based on an opinion or using only qualitative data.

In this phase we will focus on one of the easiest tools available that can be used by all levels of the work groups. From shop floor operator to the President of the company, all will feel very comfortable using this tool. Hopefully at this point in the project, the reader already has a very good concept of the 3P approach and how each P directly relates to the three levels of the organization. At this point in the project we will apply these principles to focus in on the actual problem by taking it further into each individual level of the organization to solve the problem. The main focus is to solve the problem by using a standardized approach simple enough so that all levels of the organization can understand it. In the Analyze phase we will focus on how to properly use the 3P 5 why problem solving tool.

3P 5 WHY –

Sections of the 3P 5 why are Protect, Prevent, and Predict

PRODUCT level – **P**rotect

PROCESS level – **P**revent

SYSTEM level – **P**redict

The basic principle of the 3P 5 why is to drill down to the problem by asking ourselves a series of questions for each level of the P's. All of the questions are very simple; we simply ask **why**.

Asking "Why?" numerous times may be one of the greatest learning tools a child has. If your 10 year old were to ask you to go outside and ride his bike and it is dark and raining, and you say NO, he asked:

- **Why? - Because you might get hurt.**
- **Why? - Because you could be ran over by a car.**
- **Why? - Because they might not see you in all the rain and darkness.**
- **Why? - Because your bike is not easily seen in this condition.**
- **Why? - Because you have no lights and no reflective gear.**
- **Why? - Because we have never bought you any lights and gear.**
- **Why? - Because we do not allow you to ride at night.**
- **Why? - Because you might get hurt.......**

This technique may seem childish, but why do you think a child uses this approach? BECAUSE HE IS TRYING TO SOLVE THE PROBLEM! He thinks that if he gets to the last "Why", he can implement the solution and be able to ride his bike. The reason we keep asking "why" is because if we can fully understand the actual root cause, we can implement the solution.

Always start off by first stating the problem statement. Remember from the Define phase that we must have a target, in the Measure phase we must have it quantified, and now in the Analyze phase we will determine how bad the actual situation is; and hopefully why it is bad. Always start the investigation with the Product level analysis first and then work upstream to the process and then to the system.

3P 5 WHY PROTECT –

Protect is directly related to the Product, therefore we will focus our attention to ONLY the product characteristics. We should always have the Process Map available for review and analysis in the Protect section. Form the example in the Define phase we defined a problem statement that reads: The Specification for Part **ABC123** requires the length of the part to measure **25.4mm** +/- 1.0mm. On **1/1/2009** customer **ACME** reported finding **75 pieces** from lot number **12-15-08-SAM.** The actual measurements of the parts were **22.5mm**, which is **2.0mm** out of stated specification tolerance. Starting at the Product level we would repeat this problem statement to the corrective action team and then simply ask why. We would ask: Why was the part out spec? **NOTE:** In the Define phase we have already discovered that the dimension was out of specification tolerance as compared to the target (we quantified the problem), in the Measure phase we discovered exactly how far out of spec the part was (we actually measured the problem), and now in the Analyze phase we need to find out why the part was actually out of spec. We will always start at the highest level possible (the actual problem statement) and work our way down the 5 why path. Below is a typical example of how our path might lead us for a Product level problem in the Protect section:

PROBLEM STATEMENT: ACME found 75 pcs of ABC123 that were 2.0mm out of spec.
1. Why? – Parts were shipped to customer out of spec.
2. Why? – Nonconforming parts were accepted as good parts and put in box.
3. Why? – The operator did not reject any parts during the process run of 800.

4. Why? – The width dimension was not being measured accurately.

5. Why? – The gage fixture used to measure this dimension had a loose clamp.

We will read the why, why, why, why, why on the way down, but will read **_therefore_** on the way back up the list. For example, read from the bottom of the list to the top starting with: 5. – The gage fixture used to measure this dimension had a loose clamp; THEREFORE, 4. – The width dimension was not being measured accurately, THEREFORE 3. – The operator did not reject any parts during the process run of 800, etc.... Use therefore instead of why to see if this statement makes sense.

Of course this is a very simple example but the same concept will be used every time for every situation. In the Protect section we are trying to find why we did not protect the customer from the defect. Again, from the Define and Measure phase we have already quantified the problem and measured the problem. In the Analyze phase we are trying to determine why we have failed to protect our customer from the defect; which means that we did not measure and reject the parts appropriately. We are trying to drill down to the actual root cause of why we allowed the defects to reach the customer so we can resolve it. There must be some reason causing defects to get through our process and to the customer undetected and we must find that reason in the analyze phase. We are focusing ONLY on the quality gates and our inabilities to detect defects. In many problem solving meetings many individuals will not be able to focus only on the product characteristic and will attempt to steer the team away from the subject. It is paramount to keep the team focused only on the product characteristic. There will be other opportunities in the analysis phase to discuss the process and even the system later on in the problem solving task. If the champion of the team has the appropriate material available such as the Process Map, check sheets, work instructions, and gage instructions they will be able to better focus the team to the product characteristic.

Once the team reaches the last why, read all the way through the entire list of whys to ensure the team has reached the last possible why. Does the last why describe exactly why we failed to protect the customer from the defect? If so, it will be time to move on to the corrective action to resolve the problem. If the team reaches a point where they are unable to move to another why, it will be a good indication that this is the actual problem. If the team gets to a point where they find an absence of a system, do not spend any more time in the analysis. A team could spend days trying to figure out why the system is not there rather than just fixing the problem and implementing the system. For example:

1. Why? – Parts were shipped to customer out of spec.
2. Why? – Nonconforming parts were accepted as good parts and put in box.
3. Why? – The operator did not have a gage to inspect parts with.
4. Why? – Width was being measured at the raw material state.
5. Why? – Don't know, can't determine any further.

At this point we can not determine why the process was not given a gage. We could only speculate that whoever set up the process did not feel the part needed measured again for width because the raw material width was being measured; therefore not putting a gage at the final part operation. Again, the team could spend days or even weeks trying to figure out why the gage was not being used to measure the part verses the raw material. Remember that we are only focusing on the product characteristic and not the process or the system at this stage. We may be able to answer this question later on at the system level. But, at this point in the analysis of the protect section we have determined that the gage was not being used to measure the product, only the raw material. The reason we did not protect the customer and they received parts out of spec is because we were not measuring the part in the finished stage. Now we can move to the improve phase and improve this situation so we can protect our customer in the future (Improve Phase next Chapter).

3P 5 WHY PREVENT –

Prevent is directly related to the Process, therefore we will focus our attention to ONLY the process characteristics. We should always have the process map available for review and analysis in the Prevent section. Form the example in the Define phase we defined a problem statement for the process level that reads: 75 pieces of Part ABC123 was 2.0mm out of spec (we know this from the Part level) because the **left front locating pin** in tool **XYZ789** broke during processing on **12-15-08**. The pin broke at piece part number **724** out of an 800 piece production order. **NOTE**: In the Define phase we have already discovered exactly which process the pin broke (we quantified the problem), in the Measure phase we discovered exactly when the pin broke (we actually measured the problem), and now in the Analyze phase we need to find out <u>why</u> the pin broke or why we did not prevent the pin from breaking.

Just as in the Protect section we always repeat this problem statement to the corrective action team and then simply ask why. We would ask: Why did the process allow the defect to be produced? We will always start at the highest level possible (the actual problem statement) and work our way down the 5 why path in each of the 3P sections. Below is a typical example of how our path might lead us for a process level problem in the Prevent section:

PROBLEM STATEMENT: ACME found 75 pcs of ABC123 that were 2.0mm out of spec.

1. Why? – The locating pin broke during the production run.
2. Why? – The pin was not changed out before it broke.
3. Why? – The Process Tech was not directed to change the pin.
4. Why? – The Pin has no change out schedule.
5. Why? – The Preventative Maintenance Plan did not include changing the Pin at prescribed intervals.

In the Prevent section we are focusing ONLY on the process weaknesses and the inabilities of the process to prevent the defects from being produced. It is paramount to keep the team focused only on the process characteristics. Just as with protect, there will be other opportunities in the analysis phase to discuss the system. The champion of the team should always have the appropriate materials available for the team to analyze such as the process map, process check sheets, process set-up work instructions, and cycle time requirements so they will be able to better focus on the process characteristic.

Following the exact same method as the protect section, read back from bottom to top using therefore instead of why to see if the flow actually makes sense. Does the last why describe exactly why the process failed to prevent the defect? If so, it will be time to move on to the corrective action to resolve the problem. If the team reaches a point where they are unable to move to another why, it will be a good indication that this is the actual problem. If the team gets to a point where they find an absence of a system, move on and do not spend any more time in the analysis.

Always check the work by reading from bottom to top:

PROBLEM STATEMENT: ACME found 75 pcs of ABC123 that were 2.0mm out of spec.

1. Therefore? – The locating pin broke during the production run.
2. Therefore – The pin was not changed out before it broke.
3. Therefore – The Process Tech was not directed to change the pin.
4. Therefore – The Pin has no change out schedule.
5. – The Preventative Maintenance Plan did not include changing the Pin at prescribed intervals.

Prevention section of the analysis will take some time to complete so be patient and allow the team to explore all avenues of the actual process condition. There may be more than one reason the process is not preventing the defects, and all of them should be analyzed. There are definitely many other supporting tools that can be used in the problem solving task for such situations. The team could use the fishbone analysis to find the source, thought maps, process audit results, DOE's, or even the force field analysis to pin point all possible points in the process that could fail. The 3P 5 why is a simple tool that is used as a standardized approach that anyone in the organization can understand and use. It is better to have people from all levels of the organization part of the problem solving team than to only have a few experts. The experts should of course be part of the team to help out in situations that might require more complicated analysis tools or may require some type of statistical study.

3P 5 WHY PREDICT –

The last section of the 3P 5 why in the analysis phase is Predict; which is focused on the system. This is the most important section of the entire 3P's because it will effect the entire organization. Changing the system or improving the system will have a detrimental impact on the businesses' output. Think of the system as being upstream as far as possible where the actual stream would begin. If we were to put something in the stream, eventually it would trickle down all the way to the ocean. The system is also one of the most difficult levels to change as we will learn later in the Improve phase. It is difficult because most people hate change and typical system rules have been in place for as long as the organization has been established. How difficult would it be for an organization to change something as simple as the dress code if the same dress code had been in place for 20 years? As you can see the systems level analysis will be very difficult because if people see that the problem exist at the higher levels, they will tend not to try and solve it because they are not in authority to actually make the change. For example, if an organization had a problem with taxes, the operator on the floor would do nothing because he has no control over that system. None-the-less, the system problem needs to be analyzed and the data taken to the appropriate level of manager to resolve the problem or it will directly affect the entire output of the business. That is to say, the operator may not be able to change the tax method but if the Government seizes the Company he will be directly affected.

We will use the same approach as in the Protect and Prevent sections above by always starting out with the problem statement and asking the simple question why? We have already defined the problem (quantified it), measured the problem, and now we must analyze the problem to see where the problem actually is. Let's use the same problem as above and see if we can find the systems problem by using the 5 why methodology.

PROBLEM STATEMENT: ACME found 75 pcs of ABC123 that were 2.0mm out of spec.

1. Why? – The locating pin broke during the production run.
2. Why? – No one changed the pin before it broke.
3. Why? – The locating pin was not considered for change out at prescribed intervals.
4. Why? – The PM plan was not properly reviewed by Maintenance Dept.
5. Why? – The Plant policy does not require sign off of PM plans by the Maintenance Manager.

In the Predict section we are focusing ONLY on the system weaknesses and the inabilities of the system to predict failures in the process. It is paramount to keep the team focused only on the system characteristics. The champion of the team should always have the appropriate materials available for the team to analyze such as plant policies, procedures of the organization, and tier one work instructions. If the team does not have a member of upper management during the systems analysis phase there is a very good possibility that the proper analysis will not be performed and the actual root cause will not be found. Many team members are intimidated when it comes to the system analysis because they are afraid to show upper management there may be problems in the system. If upper management rejects the analysis, at least the team knows exactly where the problem lies. It is up to the upper management team how they decide to run the business based on their own goals and objectives. In the Improvement phase we will learn how to approach those managers that reject the analysis. The team champion must encourage the members to try and analyze the system even though they might reach resistance when it comes time to implement the improvements.

Once the proper analysis has been completed, be sure to go back and see if the statements actually follow a path to the conclusion. Follow the exact same method as with protect and prevent sections. Read back from bottom to top using therefore instead of why to see if the flow actually makes sense. Does the last why describe exactly why the system failed to predict the defect? If so, it will be time to move on to the corrective action to resolve the problem. If the team reaches a point where they are unable to move to another why, it will be a good indication that this is the actual problem. If the team gets to a point where they find an absence of a system, move on and do not spend any more time in the analysis.

Always check the work by reading from bottom to top:

PROBLEM STATEMENT: ACME found 75 pcs of ABC123 that were 2.0mm out of spec.
1. Therefore? – The locating pin broke during the production run.
2. Therefore – No one changed the pin before it broke.
3. Therefore – The locating pin was not considered for change out at prescribed intervals.
4. Therefore – The PM plan was not properly reviewed by Maintenance Dept.
5. – The Plant policy does not require sign off of PM plans by the Maintenance Manager.

In the Analysis phase of this project it is important to focus solely on the section in which the team is working. If the team jumps around from one section to another it will be very difficult to analyze the data. Members of the team will quickly jump to conclusions without full analysis of the data when moving from one section to the other without completing that section first.

There is a mathematical method to consider when the analysis is being performed. It is important to note that with any change to the system, the process and the part will change as a result. Any change to the process, the part will change as a result. Be careful in the analysis and be sure to cover all possible inputs that could have a direct or indirect effect on the output. Think of the entire system when performing analysis and how the entire system was designed. The following chart illustrates the function of x and how it impacts the final output (y). When performing a problem solving project, it is a very good idea for the team to map out all the inputs (x's) in the entire process from start to finish. If the team can see where each input belongs in the entire process scheme, they may be better equipped to perform the actual analysis and how each x depends on the inputs upstream. Many times after looking at the x map, the team will be able to rule out numerous x's because of their dependence on the x upstream. In other words, if the upstream x is out of spec then the down stream x will also be out of spec. So, there is no need to perform analysis on the downstream x if this is the case. f(x)=y simply means that for every function of the input (x) the output (y) is effected.

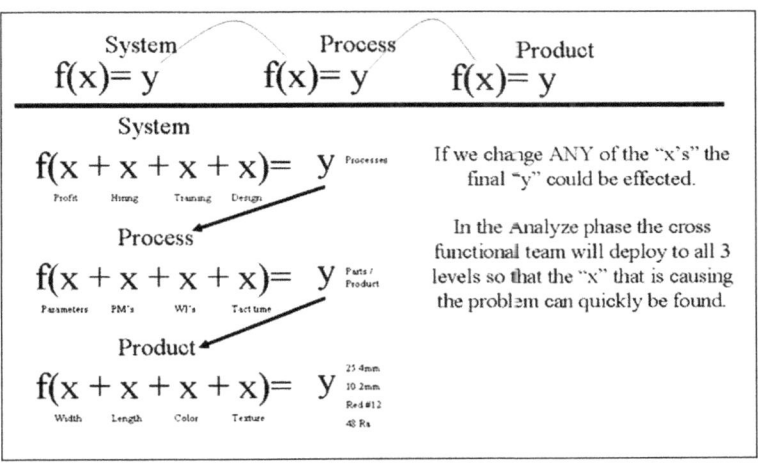

It will take practice to master this tool. The more the team uses it, the better and more efficient they become.

Measure –

Let's say for example that we were manufacturing roast beef sandwiches and the total weight of the finished sandwich had to weigh between 6 and 6.5 oz. If we made the x map we would see that the buns are the first step, meat, mayo, and then the wrapper. Let's say that the buns weighed in at 6.6 oz. Do we need to go ahead and analyze the other 3 x's downstream? It would not be necessary at this point because we already know that the bun weight exceeds the overall target goal of 6.5 oz max weight. If we had created an x map like the one below, we could quickly see at each step what the criteria was and how that criteria affects the upstream x, input. We can see from the x map below that the buns are out of spec by 4 oz (actual = 6.5 oz, spec = 2.5 oz, diff of 4 oz). Of course, if we correct the buns and the weight is still wrong we would move on to the next x and so on until we completely analyzed each x downstream.

$$f(x + x + x + x) = y \quad 6.5\,oz$$

Buns	Meat	Mayo	Wrapper
2.5 oz	3.5 oz	.25 oz	.25 oz

In the Analyze phase of the project it is important to keep the team focused on analyzing the actual data to determine why the failure occurred. Remember that problem solving is not an answer to a question; it is a solution to a puzzle. Unless we can completely understand where the failure came from without doubt we will be unable to implement the corrective actions. Analysis of the data will indicate where the part, process, or system is actually at as compared to the stated target or goal. It should also tell us why the part, process, or system is in the condition that caused the failure. There are numerous methods to analyze data with and the champion of the problem solving team should use any tool that will result in the correct analysis of the data. Use the opportunity to train others on new and innovative ways to analyze data so that in the future they too can benefit from those tools.

CHAPTER 5
IMPROVE —

At this point in the project we have defined the targets that need to be measured, we have taken measurements of those targets at the product level, process level, and the systems level. In the Analyze phase we studied the collected quantitative data to ascertain why the problem exists. In the Improve phase we will learn how to properly assign the corrective actions to those problems using a formalized approach. The most important part of the Improve phase is to implement the appropriate corrective actions at all three levels (individually) so the problems are completely resolved and will not reoccur. Improvement phase is one of the most important parts of the entire project because at this point we will decide the change that needs to be made to eliminate the problem or potential future problems at all three levels of the organization. If we fail to implement the appropriate corrective actions at any of the three levels, we could face reoccurrence of the same problem and lose the customers' faith in our abilities to control our processes and protect them from defects. At all three levels of the organization there are formalized methods to ensure corrective actions are put into place that will sustain longevity.

SYSTEM —

To appropriately improve the output we must always consider the design of the system, the process, and the product. If the process was designed to overcome the weaknesses in the system, we must go back and redesign the system. Remember, that the system controls all the processes in the organization so changing one process to make up for a poor system will only improve that one process. What will become of all the other processes that are controlled by the same system? Eventually they too will fail because of the same system weakness. Let's say that the team finds a problem with a broken pin at a stamping process and decides to create an elaborate method to prevent this from occurring again. The team spends $15,000 and invests three weeks to complete the improvement activities. They study the process and decide that the pin must be changed out every 6,500 cycles. The team revises the PM plan that requires the Process Tech to change the pin out as required. The process work instructions are revised, the Process Map is revised, and every operator that runs this process must undergo a 30 minute training class covering these new requirements. As an addition to the above, the team also installs a computer controlled program that will automatically stop the press at 6,500 cycles and send out an audible alarm indicating that it is time to change the pin. The software also locks the press start feature until the appropriate password is entered by an authorized Process Tech indicating that the pin was replaced. These improvements to this process are great and should prevent any further issues with the part because of a broken pin. What about the press sitting directly beside this one? The two processes are now very different and require different skilled operators to run them. Has making all the improvements to this process actually improved the overall system? Is there a possibility that there could be more failures at other processes within the organization because of broken pins? The answer is yes, of course there are other potential failures at any of the processes because the system has not predicted those failures either.

Stopping the improvements at the process level or trying to improve a process to overcome the weakness in the system will only result in more failures in the future at other processes. We must improve the system from the knowledge we have about this process. We know that this process has already failed and why it failed. It is only a matter of time before other processes fail because of the same weakness in the overall system. Using the same example, how could we improve the overall system so that all processes could benefit from the teams hard work and prevent future failures? One simple method would be to require all pins at all processes to be changed at 6,500 cycles. To go even further we could revise the preventative maintenance policy that requires all PM plans be reviewed and approved by the Maintenance Manager to ensure all expendable tooling be studied for change out schedules. By implementing this policy we would basically be using a single failure from one process to improve all processes. We would develop a policy that requires the study of tooling be performed up front and develop a schedule for replacement. By using this practice we could avoid tools breaking and prevent defects from being produced. We are not trying to overcome the system weakness by upgrading a single process we are actually strengthening the system to prevent future potential failures throughout the facility on all processes.

Any changes or improvements to the System will require upper management approval. Be sure to always present the improvement options to upper management before any actual actions are taken. Management is accountable for the system so changes made by those that are not accountable could result in catastrophic failures. Although the corrective action team may have a viable solution, upper management must first review all options and select what is right for the entire system based on the appropriate data. It is the responsibility of the corrective action team to present

<u>all</u> the relevant data to management so they can make the appropriate decisions that will satisfy the overall organizational goals and targets. Do not allow the team to present only the data they wish. By presenting all possible options the team can allow management to make an unbiased decision. Even if some members of the team do not totally agree, management will ultimately assume the responsibilities for the decisions.

<u>PROCESS</u> –

Many times in the Improvement phase the team will only focus on the product design. They believe that changes to the product design will overcome weaknesses in the process. In most cases that is not true. In the Measure phase we learned that the consumer actually decides the product design, so changing the design may effect the customers' satisfaction with the product. We must implement our improvements by using creative solutions to fix and prevent the problems found during the analyze phase. Solutions must be implemented at the process level by developing sound implementation plans based on a data driven systems, and doing what the data actually tells us to do. The team should create innovative solutions using technology and discipline to ensure the process is free from defects. Most importantly we must remove the intent, opportunity, & capability of the process or operator to produce a defect.

Intent - objective, unknowingly make a mistake:

"That was the way I understood it to be done." If we do not educate the operators and associates of the correct method and validate that training, we are expecting them to train themselves. If we program or set up a process wrong, enter the wrong parameters, or put in the wrong inputs, the process will respond accordingly. The process will only abide by those rules that are defined and controlled.

Opportunity – chance, favorable circumstance that allows for mistake:

"My equipment can be adjusted, so I adjusted it to finish quicker." If we do not provide the correct method and materials to perform the task, the operator or associate will select the best method that suits them and not the favorable method for the product. If we do not control the input methods into the process, the process too could use the incorrect methods and make defects.

Capability – means, actual ability to carry out the mistake:

"I had too many choices; I guess I made the wrong choice." If we give the operator or associate too many choices and allow them to decide for themselves, they may decide wrong. We need to eliminate all choices except the correct one. The same situation applies to the process. If the process can be set up to run at many different parameters and we do not control those settings, the process could also be set up wrong.

For each corrective action the team should ask; "Did we remove the intent, opportunity, and capability to make this same mistake?"

Improvements to the process level will require skilled assessments. The corrective action team may have many options but may also find that a majority of them will not work. The thought process for any improvement or change at the process level must use the win-win methodology. That simple means that the improvement must benefit everyone involved in the process in order to satisfy all the process targets. The team will quickly find that things are not as important to some as they are to others. For example; the team may decide to increase the clean up time for a particular piece of equipment to avoid build up of debris. The impact on the added time may result in fewer parts being produced which will have a negative impact on the Production Managers' daily output goal. The team may have solved one problem but now they have created another problem. The change could also have a negative impact on overall profit which would result in an even bigger problem. Improvements to the process are not easy because of the local impact they may have on the entire process environment. The team must use the tools they have available and look at all possible outcomes to any improvement before it is made. The team must compare the actual result of the change to the actual result of the output. Always try and assign a dollar figure to each improvement so management can use that data to make the appropriate decision. Use technology and discipline to attack the problem and try to avoid additional work being added to the process. A process with fewer steps has a far greater possibility to succeed because there are fewer steps that could go wrong. A change that results in profit loss will be almost impossible to get approved as compared to a cost savings change.

Any changes or improvements to the process will require the process owners' approval. Be sure to always present the improvement options to the owner before any actual actions are taken. Hopefully, the corrective action team has included the process owner as part of the team so that any changes or improvements can be made with their input. The process owner is accountable, so any changes made by the team without their approval could result in catastrophic failures. As with the system, the corrective action team must present <u>all</u> relevant data so the appropriate decisions can be made that will satisfy the overall process goals and targets. The team may have short term and long term solutions to present to the process owner so always present all options for each.

PRODUCT –

Improvements to the product should be the easiest level of the Improvement phase and will require a decision at the lowest level. In most cases it is a matter of putting the part or product back to the defined specification. The operator or associate can make improvements at the local level to correct the part. Operators are commonly given the ability to make slight adjustments to the process to ensure the parts are produced to specification. In most cases the improvement to the system and to the process will automatically correct the part defect. The improvement on the part level may include some type of detection method that may have not been present at the time of failure. The team may decide to add an inspection gate, a new or different measuring tool, or increased inspections to detect future failures. The corrective action team champion should have the authority to make necessary improvements at the part level without upper management approval. Keep in mind the win-win methodology at the part level as well. If the team increases the inspection time it could result in a production or profit goal to fail. The corrective action team should be made up of individuals that can adequately make improvement decisions at the part level without involving higher level managers.

Typically, most corrective action teams will implement a temporary containment plan to ensure 100% inspection of the parts until corrective actions are implemented at the process and system level. Keep in mind that 100% containment is an unexpected expense by the organization and profit goals will almost always be affected. Containment is a very good data collection opportunity because every part is being inspected and defects can be found as they are produced. It should never be designed into the process long term because of the reliability and dependence. For example, if an operator producing a part knows that someone down stream is 100% inspecting their work, they tend not to take ownership of their process. They figure that if they miss something that the person down stream will catch it. The person conducting the containment down stream is also thinking that the person producing the part upstream is inspecting the part and should find any defect therefore they do not need to inspect that closely. A good containment will be controlled to a point that the process operator will be made aware of any defect found down stream and will be notified immediately when defects are found. The containment is more or less a grading of the process and the operators ability to detect defects. If either fail, the process operator should be made aware so corrective actions can be implemented in the detection of defects. Many times in the measure phase the team measures the process as it exist and finds no problem. Containment may be the only method that allows them to study the process for a longer period of time to actually witness any change points while protecting the customer at the same time.

If the improvement to the part requires a design change, it must be presented to upper management to decide. Careful consideration must be made to any design change because of the detrimental effect it may have in the market or to the process that produces it. The design change could totally change the processing method all together which could result in even bigger unknown problems. Minor design changes may seem insignificant but could result in customer dissatisfaction in the market. For example; a corrective action team decided to remove a paint drain hole in a grip handle that goes on a golf push cart. The hole had a tendency to shift during the manufacturing process, so eliminating it seemed the best solution to resolve the problem once and for all.

The drain hole was no longer needed anyway because of an improvement made in the painting process. Once the customers started receiving the push carts without the drain hole they started complaining and refused to purchase any more carts. The corrective action team did not realize that the customers were using the drain hole to mount a small map of the golf courses in which they were used. Even a small change in the design could have major impact in the market. Worse case the customer could be injured by such changes and the reputation of the organization could be shattered. Design change may be the only option, but all other options should be considered by the team before presenting them to management for review.

Any improvements to the product should always include some type of detection. One of the best ways to ensure the process is running as it should is to validate the output. Although some processes may run as close to perfect as possible, there should always be some type of inspection on the product to ensure it is running as intended. Long term data should always be available so that any changes in the process can easily be detected and problems prevented. The only way to collect the long term data is to implement some type of inspection gate at prescribed intervals. Build into the process automation that can detect part failures when possible. Give the process operators ownership and allow them to make the correct decisions. Take away the intent, opportunity, and capability of the operator to make a mistake. Do not become dependants on down stream inspections; only allow conforming parts to leave to the next step in the process.

CHAPTER 6
CONTROL –

At this point in the project we have defined the targets that need to be measured and taken measurements of those targets at the product level, process level, and the systems level. In the Analyze phase we studied the collected quantitative data to ascertain why the problem exists. In the Improve phase we learned how to properly assign the corrective actions to those problems using a formalized approach. In the Control phase we will learn how to appropriately ensure the improvements are maintained and readjust the targets to ensure on-going improvements are maintained.

The Control phase of the project is not the last step. It is the last phase in the problem solving project, but a new beginning to a better system, process, and part. In the Control phase we will learn how to maintain those improvements that were made on the system, process, and the product and how to continuously improve them. In order to ensure those improvements sustain longevity we must implement controls so that we consistently monitor each improvement made. We must continue to adjust the targets so that we continually improve year after year.

Any improvements that were made during the problem solving project that have no controls will eventually fail again. In order that we do not suffer a reoccurrence of the same problem we must ensure the improvements continue for the life of the system, process, and product. Most problems that we solve are problems that have been solved previously but somehow somewhere the controls failed to maintain the improvements. Almost always a reoccurrence is the result of a target that is either never established or never monitored for improvement. If we all had a dollar for every time we heard the statement, "We used to do it like that, and never had a problem" we would all be rich! We tend to fall back into our old ways of doing things and end up back at the same problem we started with because we fail to monitor the target and maintain control of the improvements.

A good control method is a system that constantly measures the improvement. We must continue to measure the improvement at all three levels of the organization to ensure the desired output is being achieved. We should never settle with the good enough attitudes because it leads us to believe that our system is free from problems. We should always look at the targets and adjust them to achieve perfection. If we continuously work on the targets to improve them we will become experts at the problem solving task. Just like an Olympic runner training for the 100 meter sprint, he or she is never satisfied with the results of their time. They know that if improvements are not made that eventually an even faster runner will come along and take the medal. The more we use and practice the problem solving tools the better we become at preventing problems before they occur.

Most companies that suffer from numerous problems tend to work on the protection side of the equation to often and very little on the prevention side. In other words, they wait until a problem occurs verses preventing the problems from occurring in the first place. Always challenge the organization to see the cost factor for protection verses prevention. Let's look at the broken pin issue one more time. If we had predicted the problem by first studying the tooling components and replaced the pin at the appropriate cycle, we would have avoided two major aspects. The first is obvious; we would have avoided all the containment cost, replacement part cost, shipping new parts to the customer cost, and manpower cost from having to reproduce the parts again. The second aspect may not be so obvious, but is the most important. What about the customers' faith in our abilities to produce good parts? They pay for good parts, base their performance on our abilities to provide good parts, and yet we cannot provide that service for them. If we continue to have failures the customer will no doubt find someone that can give them what they pay for.

A twenty-two year retired auto salesperson once told me that at his most profitable job he had sold over 150 cars in a single year. He was working at a newly opened dealership when it occurred and the dealership was well on their way to becoming very successful. He was elated to be a part of the growing team. The dealership had made many improvements from the previous year to combat some of the problems that had occurred during their growing year. One of the improvements made was adding a very detailed check sheet so that the salespersons selling a vehicle were sure to offer the customers every available option offered by the dealership. The second year was off to an even bigger sales volume for my friend the salesman and the dealership when he ran into a problem. One of his customers' purchased a very expensive used work truck that soon blew an engine just a couple of months after he purchased it. Admittedly, he was so busy during the sales transaction that he forgot to use the mandatory checklist and go through each option with the customer. Item number 2 on the checklist was an option for the customer to purchase an extended warranty that would cover the drive train for an additional 12 months or 12,000 miles. Because of this, the customer was not aware of any such warranty and did not take advantage of it. The truck unfortunately blew the engine and left the customer stranded at his place of work. His job site just happened to be a road construction project for a new freeway off ramp that leads right to the dealership where he purchased the truck. My friend was sympathetic with the customer but according to the store policy could not replace the engine or even cover a portion of the replacement cost because there was no extended warranty purchased. Of course the customer argued that the warranty was never offered to him and if it had been he would have purchased it, so the dealership was at fault. The dealership refused to admit fault and the customer was left to pay for the repairs himself. Out of over 150 completely satisfied customers the previous year the salesman never had one person to recognize him; until now. The disgruntle customer was determined to let everyone know about his situation so that no one else would have to suffer his injustices. He simply pushed the massive white truck just to the edge of the off ramp and painted a large yellow lemon on the side. He followed his art work up with a bold statement painted in bright safety orange that could be seen for a mile up the interstate from both the north and south bound lanes. In 3 foot high letters he wrote on the side of the truck: FOR A REAL LEMON LIKE THIS ONE, SEE (my friends' name) AT (dealerships' name). He also painted a 10 foot horizontal arrow on the truck pointing the way to the dealership just at the end of the new off ramp. Needless to say that my friend did not have another sales breaking quota level as he did the previous year; nor did the dealership.

Hearing this story and actually seeing the truck with the big yellow lemon painted on the side, I had no choice but to try and help my friend solve this problem. My friend and I formed a small corrective action team complied of several salespersons along with the General Manager and set out to implement some improvements to avoid this type of future publicity. Using the same approach as described in this project we found the following (brief version):

	DEFINE	MEASURE	ANALYZE	IMPROVE	CONTROL
PRODUCT	Warranty not sold to customer	Checklist was not present in the customers sales file to ensure options were offered	Customer file had no checklist, but had all other sales documents (4 total)	Include Manager sign off requirement for all documents including checklist before car is sold	Finance Officer can not enter file unless checklist is part of sales agreement (no money can change hands).
PROCESS	Checklist is a separate document from sales agreement, dependant on salesperson to complete	Salesperson has 5 documents to complete for each customer sale. Salesperson only has access to 3 documents from their computer	All 5 documents are printed from separate files in separate locations that have separate authorizations	Combine all 5 documents into 1 common file. When printing, all 5 documents print at once at the Salespersons desk	Documents are coded by page number, last page has signature boxes for all approvals of sale
SYSTEM	Customer records were not being verified when checklist was added as a new requirement to ensure checklist was being used	110 checklist were completed as compared to 250 customer records audited by management. Only 44% records had the checklist	Manager was not required to check customer files to ensure the checklist were being completed by salespersons	Revise Managers job description that requires inspection of customer files to ensure checklist & all other documents are present	Manager must report to CEO each week with a status report of records completion compared to cars sold (goal 100%)

Ensuring that we resolve this problem and prevent a reoccurrence, we covered all three levels of the organization with the problem solving project. As seen from the brief example of the project there were improvements made at all three levels. If we had only improved the product level, there could have been future opportunities for failure from the process and the system. By also including the process and the system during the project we were able to implement improvements that not only prevented future problems, it also increased productivity of the salespersons. During the analysis it was discovered that none of the salespersons had access to the documents that they needed to complete an approved sale. Had we not included all three levels of this project, most likely there would have been another failure for another missing document in the near future. We also found that when the original improvement was made to add the checklist there were no controls in place to ensure the change took effect. The checklist was a great improvement but was only being used 44% of the time. Any improvement made at any level in the organization is basically a ticking time bomb unless it is followed up with controls to ensure its' longevity. The original idea for the checklist was implemented because of a similar problem the dealership had the previous year but obviously no controls were put into place to ensure it was being used. Later, after the dealership completed this project they actually went back and inspected all 1,650 sales file records and found an average of only 20% of the files had all required documents. Why? No one was assigned accountability to ensure the files were correct.

ACCOUNTABILITY –

Accountability is one of the major factors of a good control method. Someone or some entity must be accountable for any target that is established. Let's say for example that my wife and I give our son a curfew to be home by 10:30 pm on the weekends. If we do not decide who or how we will be monitoring this target how will we know the target is being achieved? When we found out that he had been arriving home late for several weekends and confronted him about the discrepancy he simple told us he thought it was OK because no one had scolded him the first time it occurred. We implemented a target goal and failed to implement the controls to ensure it was maintained. He was under the impression that the target was not being monitored so it was merely a suggestion to be home by 10:30 and not a rule. Controls must be considered with due diligence when put into place because the person that is accountable must be willing to accept the new task. In the above example of the dealership, when controls were put into place the Manager had much more work to do than he did before. He had to arrange his previous schedule of work to ensure he had time to check all the documents, but by being proactive eventually his reactive task dwindled so he had even more time on his hands to make further improvements. Controls do not necessarily have to be assigned to one particular person. They can be assigned to a group, department, or even a piece of equipment. My wife and I decided to simply let the home security system control our son's curfew. We programmed the security system to activate promptly at 10:30 pm. Any persons entering or leaving the house through any opening after such time were quickly identified as intruders (or some one that broke a rule). By implementing the controls we are letting our employees know that this rule is important and must be maintained. We are also letting them know that this rule will always be monitored by whom or what.

Controls are not only about ensuring a target is achieved or that an improvement is being maintained, it is about continuous improvement. As we find problems in the organization and implement improvements, we are in a sense adjusting or creating new targets to achieve. By constantly adjusting the targets as a result of an improvement made, we are aligning the organization to be more profitable and more efficient to accomplish the overall mission. Controls allow us to monitor the improvements to ensure we are successful and that each upstream input is achieving its' target as intended. We need to always remember that each level of the organization must have controls in place to achieve the overall output. Product controls ensure the customer is satisfied. Process controls ensure the part meets specifications. System controls ensure the organization sustains longevity and continues with their business scope.

Project Checklist:

Define Phase	1	Form the corrective action team	☐ complete
	2	Quantify - narrow down the problem	☐ complete
	3	State the problem (write problem statement)	☐ complete
	4	Decide if outputs are actually measureable (quantifiable)	☐ complete
	5	Gather part specs, process parameters, system policies	☐ complete
	6	Compare problem to actual quantifiable specs	☐ complete
	7	Identify what tool was used to quantify the problem	☐ complete
	8	Assign responsibilities to each level of the organization	☐ complete
	9	Implement protection tools to protect the customer	☐ complete
Measure Phase	10	Identify Metric (measure of project/process performance)	☐ complete
	11	Process/Product Map - perform gap audit	☐ complete
	12	Verify the Gauging method used to measure output	☐ complete
	13	Complete YX Diagram - Identify the X's	☐ complete
	14	Complete problem prediction list	☐ complete
	15	Determine Current Capability of the process	☐ complete
	16	Complete Project/Cost Benefit Analysis	☐ complete
	17	Prepare Define / Measure Report Out	☐ complete
Analyze - Phase	18	Complete Distribution Analysis on Y (fishbone)	☐ complete
	19	Based on results perform 5 y analysis - on system	☐ complete
	20	Develop problem solving teams to address systems issues	☐ complete
	21	Eliminate obvious noise in the process (special causes)	☐ complete
	22	Complete Analysis Planning Sheet	☐ complete
	23	Assign accountabilities to maintain target goals	☐ complete
	24	Assign members to conduct sample testing analysis	☐ complete
	25	Prepare Analyze Report Out	☐ complete
Improve - Phase	26	Present Improvement to Management for approval to change	☐ complete
	27	Based on results implement formal change	☐ complete
	28	Develop contingencies to cover special causes	☐ complete
	29	Train associates to new or revised procedures	☐ complete
	30	Prepare Improve Report Out	☐ complete
Control - Phase	31	Assign Accountabilities to new improvements	☐ complete
	32	Implement verification plan to measure longevity of improvement	☐ complete
	33	Prepare Final Report	☐ complete

CHAPTER 7
CONTINUOUS
IMPROVEMENT –

Regardless how good for how long we have been to our customers they still expect to see improvements within our organization. It is very common for suppliers to request cost downs from their suppliers on a periodic basis. Do you ever wonder why customers request cost downs? Cost downs are a result of the consumer market always demanding goods and services to be better and cheaper. Consumers always want more for less, or at least more for the same price. Would you spend $1,100 for a DVD player? The DVD players just 5 years ago are far inferior to those produced today but are only a fraction of the past cost. The electronics market is a perfect example of a group that continuously improves the product, the process, and the systems of manufacturing electronic devices to remain competitive. How long would a major electronics organization last if they failed to continuously improve? It would be safe to say they would not last very long because a business is difficult to sustain without revenue.

A close friend of mine started a lawn care business many years ago with the minimum equipment. He had a small riding non-commercial lawn mower, a trimmer, edger, and a few hand tools such as racks and shovels. His equipment was what a typical home owner would purchase to maintain their own lawns. He also used his old personal truck to carry all his equipment from job to job. Nothing fancy, but he did an excellent job at a good price and had a great reputation around town. He had clients in some of the priciest neighborhoods around town. He was very profitable with his small business the first couple of years and started to add more and more customers to his client list. Before long, he had five to six lawns to mow Monday through Friday making over $10,000 per season. Of course as the business grew his non-commercial equipment could not take the abuse and it started to break down almost on a daily basis. The truck and the equipment started showing their age and it was evident that he was not in the same league as the larger commercial lawn care businesses. He never missed a job and managed to keep everything running and caught up even if he had to work on the weekends to do it. He only increased the cost by 5% each year to cover inflation and always got the job done no matter what. He was one of the hardest workers I have ever seen to this day. It seemed that he had an ever lasting business and devoted clients that he could count on for regular seasonal revenue. He unfortunately had to cease business at the end of the third year after start up. He missed one very important aspect of owning and maintaining a business; continuous improvement. Other lawn care companies that started up about the same time within the area were not only offering lawn mowing services, they also offered tree trimming, landscaping, and minor outside maintenance. The price to mow a lawn was much cheaper from the larger lawn care companies because it could be done in about half the time with their larger commercial lawn mowers. They would arrive at the million dollar homes in sparkling new trucks with a plethora of commercial grade equipment at their disposal. They were in and out within minutes it seemed compared to my friend having to take up to one hour for the same lawn. One-by-one his long devoted customers stopped calling and his business continued to fall to a point it was not worth it for him to even load up the equipment.

The first couple of years were great for him and everything he thought was well on the way to sustain the business. A business that fails to understand the customers' needs of the future will never last. There will always be another business that is willing to give the customer more for less and will have no problem taking your clients from you. Even though my friend was not interested in building his business to a larger scale or offering other services, he failed to continuously improve what he had. He could have invested in newer more efficient equipment. Even if he did not give the customers a cost down he could have done the same job quicker and maybe offered to edge the yard for free. The bottom line is the other lawn care businesses continued to improve and offer their customers more for less and eventually took over the business. The business does not necessarily have to grow to be competitive, but it will have to improve to show the customer that they intend on continuing the business and its' products or services.

ADJUSTING THE TARGETS AND GOALS –

Continuous improvement is a matter of always adjusting the goal toward perfection, lower cost, and faster production cycles; more or less it is customer satisfaction. Remember that the customer may be internal or external. Any process downstream from the system would be considered a customer, and any process upstream from the end user would be a supplier. Goals should be monitored as described in the Control phase of this project, but consideration should always be given to improving those goals on a regular basis. Many years ago while attending a management review meeting we were going over each of the organizational goals and targets for the previous year. As we discussed each goal the

Plant Manager did not seem at all interested in improving the goals for the next year. I simply asked the question as to why the goals were not being adjusted. Why was the bar not being raised? He sat with a very confused puzzled look on his face and made the following statement to the management team, "We made it just fine last year and hit every target that was given to us. Let's not set a target that we can not achieve for next year. We know we have the abilities to maintain the current goals based on our past performance, and we intend to do the same thing again next year." I quickly had flash backs of all the things we did wrong the previous year and was horrified that we would have to experience it all over again. That was his words as I understood it and we will be doing the same things we already did. Well, it didn't take a rocket scientist to figure out that we did exactly the same things we did the previous year and our customers' really did notice. In fact, they pointed it out to the CEO every time we had a reoccurrence of the same problem. Within the next year, it was agreed upon by the NEW Plant Manager that the next years goals and targets would be adjusted and we would improve our business. He along with the remainder of the management group could see a need for continuous improvement.

It is a very simple process to implement a continuous improvement plan. There are only a few steps that need to be followed to ensure a plan is in place and maintained. It must be part of every problem solving task that is performed.

1. Develop an initial target (define it in a quantifiable manner).
2. Measure the actual output as it currently exists.
3. Analyze the data to determine why the output is performing like it is.
4. Improve the part, process, or system to be better than the current output performance.

5. Control the improvement to ensure it does not revert back to the old target.

6. Continually adjust the target to force the team to analyze the outputs.

Every time a team solves a problem, they should be adjusting the current target or creating a new target. Each target should be assigned to either a person or a department to report and maintain. The target performance should be reported to upper management on a regular basis and upper management should continue to challenge those outputs. My friend that had the lawn care business learned a very valuable lesson about continuous improvement a little too late. After studying the situation and performing a problem solving project of where he went wrong he was able to start over. He first outlined his organizational goals of what we wanted to do with the business. He included improvements that he would make to the business each year so it would sustain longevity, remain competitive, and show customers that he would be around for a long while. He first invested in newer equipment and decided that he would set money aside from each job to invest back into the business to keep the equipment upgraded. He targeted larger more complicated jobs that the larger lawn care businesses did not want because of the extensive labor involved. He offered returning clients a 10% discount from the pervious year and would subcontract out any other work the client requested that he could not perform. He refused to tell the customer "No" I cannot do that. He took on jobs that did not require long term contracts as did the larger lawn care services. He would maintain lawns for those that went on vacations as a short term deal. Within the fifth year of getting back into the business with his new found aspects on continuous improvements, he was making in excess of $250,000 per year. He was the major provider of lawn care services in the area. I seen the goals list he had made some years back. It was refreshing to see where he had scratched out previous targets and adjusted them to be more aggressive each year. One particular target I found interested on the original list he had a goal that had been marked through and adjusted that read: ~~be competitive with XYZ Lawn Care within 5 years~~. The new goal read: Purchase XYZ Lawn Care and take over their contracts. By the way, he did accomplish that goal and is still doing very well today with over 90% of the business within his area (XYZ is no more).

Below is a simple illustration of a typical continuous improvement loop.

1. Problem is found
2. Solution is implemented
3. Controls are put into place to maintain improvement
4. Target is readjusted to continuously improve

CHAPTER 8
PROCESS MAPPING –

A process map can be very simple or it can be very difficult depending on the application in which it is used. A process map is basically a graphical representation of how the process or the system is actually set up to work. It must include every step in the process. The benefit of using the process map is to ensure each **x** or input is considered. As stated in the Measure phase, the process map can be very beneficial while performing problem solving projects. It should illustrate every possible opportunity to measure the part, process, or system. Many hours of investigational work can be avoided if the team is able to use this map to locate such collections and control points. A typical process map should include:

- Both value and non-value added activities
- All inputs and outputs at each step in the process
- The parameters, settings, specifications, controls of each step
- Quality Gates (inspection points)
- Performance goals for each step (scrap rate, rework rate, cycle time)
- Owners of each process (who is accountable for the process step)

As mentioned in the Measure phase, it is typical to find hidden factories that are not listed on the process map. A hidden factory is an operation within the process that is being performed and is not part of the normal process design. When developing the process map it is imperative that the creators include the actual process owners so they can participate in the development. They will be able to describe the actual process in detail including any hidden factories that may have not been considered by the designer. The core team that is needed to create a process map should include Design, Manufacturing, Quality, Operations, and the upstream/downstream customers. Always start at the beginning when creating a process map. Go as far upstream as possible and simply document each step in the process and the control methods that are used to ensure the characteristics are maintained. The following illustration demonstrates an example of a simple process flow map.

PROCESS MAP

Output Description: Bracket 1
Part Number: ABC123
ORIGINAL DATE: 01/12/07
REVISED DATE: 12/15/08

STEP	OPERATION DESCRIPTION	Cycle Time	Value or non-value added	PRODUCT OR PROCESS CHARACTERISTICS	CONTROL METHODS
1	Receive & Inspect Material	5 minutes per coil	Non-value	Weight of coil spec 800 lbs +/- 10 lbs	receiving Inspection sheet FORM-4557
				Width of coil - 50mm +/- .5mm	receiving Inspection sheet FORM-4557
				Material Certifications - Present	Chem Report DEC 1010 Rev 4
2	Move to Warehouse	1.5 minutes	Non-value	Prevent damage by using block skids to move	Inspection by warehouse tech, logged on FORM-8910
3	Move to Press 5	2.2 minutes	Non-value	Prevent damage by using block skids to move	Inspection by Press Operator, logged on FORM-8910
4	Load Coil on Press	12 minutes	Non-value	Coil must turn counter clock-wise	Coil receiver controls only allows loading from 1 side
5	Inspect Coil for damage	3 minutes	Non-value	No obvious damage to edges of coil	Visual inspection per criteria APP-3212
6	Press Bracket - Inspection	30 seconds	Value	Width of coil - 50mm +/- .5mm	Inspection sheet FORM-2112
				Length, width, standoff location	Calibrated check fixture GAGE-ABC123
7	Package Parts	3 seconds per part	Value	100 pcs per box	Box ALPHA445 has separators for exactly 100 pcs
8	Final Inspection	2 seconds	Non-value	Weight of parts box = 50 lbs +/- .10 lbs	Conveyer scale automatically weighs each box for conformance
9	Move to Warehouse	1.5 minutes	Non-value	Prevent damage to boxes by using designated aisle ways and designated warehouse locations	Warehouse software automatically selects correct storage shelf.
10	Load on Truck	4 minutes per skid	Value	Use pallet jack and individual skids to prevent damage	Truck driver performs box inspection as secondary inspection, FRM-PCK3

As seen in this example, each step in the process is documented in the correct ascending order from start to finish. Each step has a description describing what the process step is in simple terms. Each inspection point is documented indicating what characteristic must be verified at that step including the specifications for each characteristic. Each control method is listed so that anyone can see where to find the collected information if needed.

Let's say that we had a complaint from a customer that they received a box of parts and the quantity was incorrect. Of course, we asked the customer to provide us with the qualitative data so that we may actually define the problem. The customer responds: On 12/14/08 we received a box of part number ABC123 that only had 95 pieces in the box. The order requirement was for 100 pieces. The spec requires each box to have 100 pieces but the customer only received 95. Use the DMAIC approach and the process map to solve this problem.

DEFINE – Customer only received 95 pieces out of 100 on 12/14/08 of part number ABC123.

MEASURE – Checked current process and no records are being collected for individual weights.

ANALYZE – Step 8 in the process does not collect actual weights, it only accepts or rejects.

IMPROVE – Revise conveyer scale to actually output the actual weight of each box.

CONTROL – Collected data must be analyzed by QA on daily basis and posted on info board.

Very quickly we could see a problem with the final inspection of part count (weight count). The equipment only accepted or rejected finished product and did not provide us with the actual weight for each box. It is very difficult to analyze this because the data was not being collected. Although in the measure phase we should actually set up a measurement system and study the output of the line, for this example we will assume we did. We can see that the controls in step 8 need to be improved so that we can see the actual weight counts from each box. Using our formal change process we revise the Process Map to the following:

8	Final Inspection	2 seconds	Non-value	Weight of parts box = 50 lbs +/- .10 lbs	Automated Scale - Data stored in software - over under condition shuts and locks conveyer

Process Maps should be living documents. That simply means that they should be updated frequently anytime changes and improvements are made. The process map should be the most up-to-date document that is part of the process. The more the team uses the process map, the more defined and the more detail it will have. In many cases the team can solve the problem in a fraction of the time by studying the process map and looking for sources of variation. It can be designed in almost any configurations that the team can imagine. Some organizations use flow charts with actual pictures of each step as a more visual approach, while others use more simple maps as the one in the example above.

AUDITING –

The actual process should be audited on a regular basis using the map as the guide. The organization should assign individuals and teams to regularly walk the process using the map to ensure each step listed on the map is actually being performed as indicated. As part of the continuous improvement plan within the organization regular audits should be performed to help prevent problems before they occur. Layered audits are a typical approach to ensuring that all layers of the organization are also participating in the audit process. Floor supervisors all the way to the top managers should participate in the audits. Schedule these audits so that each layer of the organization participates at prescribed intervals. An example of a typical audit plan is illustrated below.

Audit Responsibility Matrix

Process	Line Leaders	Supervisors	Quality	Plant Manager	President
Receiving	Daily	Weekly	Weekly	Monthly	Yearly
Press 1	Daily	Weekly	Weekly	Monthly	Yearly
Press 2	Daily	Weekly	Weekly	Monthly	Yearly
Press 3	Daily	Weekly	Weekly	Monthly	Yearly
Warehouse	Daily	Weekly	Weekly	Monthly	Yearly
Shipping Dock	Daily	Weekly	Weekly	Monthly	Yearly
Assembly line 1	Daily	Weekly	Weekly	Monthly	Yearly
Assembly line 2	Daily	Weekly	Weekly	Monthly	Yearly
Paint line	Daily	Weekly	Weekly	Monthly	Yearly

The organization should create an audit check sheet to ensure each layer of management is participating in the audits. The audit check sheet should also include a section to annotate the findings and failures that were discovered during the audit. The results of the audit should be reported to management and the appropriate corrective action should be implemented. One of the best starting points to start a continuous improvement plan is through the use of audits. The failures found during an audit should always be included into the corrective action system.

Just as with our problem solving activities, we should always include all three levels of the organization when performing audits as well. Audits are basically taking witness of the actual condition and comparing it to the known standards or specifications. There are three specific types of audits that should be performed within the organization.

1. Product level audit – Performed at the part level. Actual product characteristics are being evaluated such as dimensional, functional, endurance, capability, part to part variation, or performance output. Sample sizes should be considered when performing a product level audit so that all possible variables are evaluated. For example, if we were performing a product audit on a bracket that is produced on multiple shifts by multiple operators we should consider increasing the sample size to include more than one shift and more than one operator. We obviously can not measure every part from every operator from every shift. That is why it is called an audit; it is only a sample size and not the entire population. We should also consider other inputs that might change such as material lots, start-up warm-up periods, and the actual environment (temp and time of day). To perform the audit we pull samples and measure them against the known standards. It is important to

also evaluate part-to-part, operator-to-operator, and shift-to-shift variations. Commonly, we find all parts within spec but see a difference between operators or different set-ups. This is an indication that the process has the potential to produce variation from such changes and is in need of better process controls to eliminate the variation.

2. Process level audit – Performed at the process level. Actual process characteristics are being evaluated such as parameter settings, equipment controls, operator training, speed and feed of process. Any item that is listed on the process map should be evaluated. Also consider the sample size when performing a process audit. Obviously not all processes can be evaluated so the schedule should be made so that at least all processes are evaluated within a prescribed timeline. The process audit should be conducted during different shifts with different operators. Always consider the actual output of the process and what the process is actually producing when developing the process audit. Consider the scrap rate, down time, rework rate, and cycle time performance. Looking at the actual output of the process will allow for a better more defined process audit. Remember that the audit is looking at the actual output as compared to the current goals and targets. The audit should indicate the gap between the goals and the actual outputs.

3. Systems level audit – Performed at the systems level. Actual systems outputs should be evaluated such as turn over rates, profit, new product development, quality performance, process map development, organizational work instructions, process design, and production performance. Typically, most organizations will have some type of industry recognized

standard that they are members of and are audited to on a periodical basis. It is always a good practice to perform these audits internally before the paid auditors arrive. Include the systems audits as part of the normal layered audit process. Assign certified internal auditors to perform the systems audits on a regular basis. Portions of the system should always be audited as part of the continuous improvement plan each time a problem solving activity is completed. For example, if the corrective action team found a problem with the operator training plan during a problem solving project, the team should audit the current policies and procedures to the actual condition that is occurring in the company.

CHAPTER 9
YX DIAGRAM TOOLS –

In order to help drive the problem solving team with projects and track continuous improvement activities, the organization may choose to use a very simple tool like the YX diagram. The diagram is used as a systematic approach to solving problems in the correct order. By using the diagram the team can see what problems need to be solved upstream that are the root cause of the downstream failures. For example, if we had a problem with an operator not being able to comprehend work instructions for a specific piece of equipment because of his educational level we would need to look upstream to solve the problem. We could spend hours and hours training the operator how to read to solve the problem on the process level, but it would not solve the systems level problem. The system should not allow operators that can not read to be hired by the organization in order to solve the actual root cause of the problem downstream. If we continue to ignore the system failure we can expect for this to occur again and again. It would do no good to try and solve the problem at the process level if the upstream input continues to allow such inputs into the process.

Think of the YX diagram like a map of an actual river. There would be many smaller streams (inputs) running into the river to form the actual river itself (output). If we tested the water and found that there was some type of contamination in it we could do one of two things: 1. we could filter the water to make it clean, or 2. we could go upstream and find where the contamination is occurring and stop it. Most organizations are set on trying to just filter the water because it seems like the easy thing to do to ensure the water is clean. The cost associated with filtering will be much-much greater than actually going upstream and stopping the contamination all together. The point is, no matter what we do downstream to change the output we have done nothing to actually prevent it from occurring again in the future. If we do not properly evaluate candidates for reading skills before we hire them how can we expect them to read the work instructions at the actual process? The problem solving team will find the YX diagram beneficial when it comes time to map out the actual input failures because it is a graphical illustration of all the inputs that need to be considered in order to correct the output. Below is a simple example of an YX diagram that can be used to map out the problem inputs

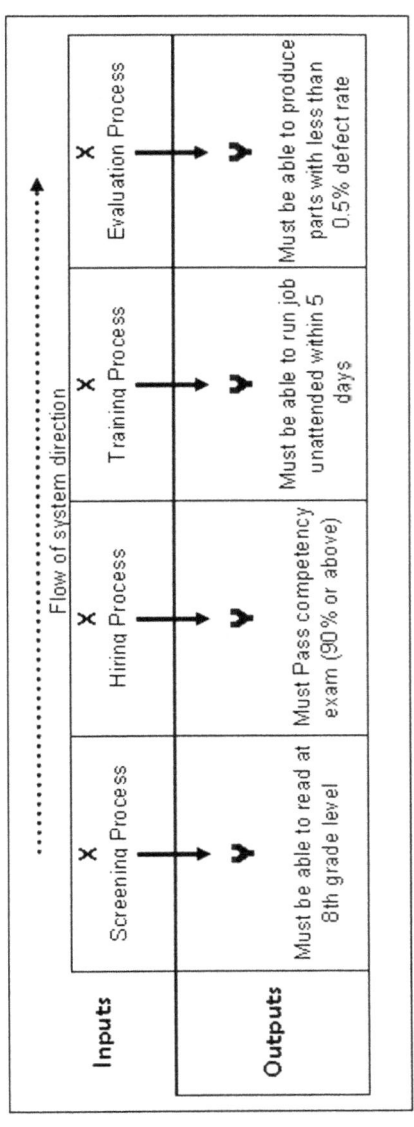

In the illustration above, if we have a poor screening process and allow employees to be hired that can not meet the goal (must have 8th grade reading level), we should not expect them to be able to meet any of the downstream goals either. If they can not read at the 8th grade level how will they be able to run the job unattended within 5 days (the downstream goal)? It would not be possible.

The YX diagram is very similar to the process map explained in the previous chapter except it is more focused on the system level inputs. It could be used for the process level or even the part level inputs as needed. It can be as elaborate or as simple as needed and can be used in many different formats. Below is another example of a more elaborate YX diagram that includes the entire DMAIC road map. The team may elect to use a similar diagram so that the input that is being evaluated can be studied all the way through the corrective action process. Using this format will allow the team to report to management much easier because it shows all upstream and downstream inputs, the goals of those inputs, why the goal was not reach, what improvements will be made, and what controls will be put into place to maintain the improvements. NOTE: Read the YX diagram from left (upstream) to right (downstream).

YX Diagram Example

Flow of system direction →

	Inputs	Preventative Maintenance Schedule	Equipment Down time	Production Goal	Product On-Time Delivery
Define:		X	X	X	X
	Outputs	Must complete 5 tune up's per week	Less than 10%	Must be 98% or higher	99% or higher
Measure:	Actual Results:	**Completed 4**	**Actual 15%**	**Actual 97%**	**Actual 96%**
Analyze:	Why was the goal not met?	Maintenance worker called in sick on Friday	Machine broke down on Friday evening because no tune up performed	See upstream failures	See upstream failures
Improve:	What was changed to correct the failure?	Assign back up worker to cover in case workers call in sick	If tune up is not performed, operator alerts supervisor		
Control:	How will the improvement be monitored?	A matrix of back up workers created and posted in all work cells	Alert light installed at process as visual warning		

In the above illustration (read from top to bottom, left to right) we can see that there was an upstream failure to meet a goal which had a direct impact on all the downstream inputs. The Preventative Maintenance was supposed to be performed on the equipment 1 time per day but was not performed on Friday. Because it was not performed we can see that all downstream targets were affected and did not meet their goals as a result. By using this diagram we can quickly go to the upstream input that caused the contamination in the stream and remove the problem. The above example of the YX diagram also displays why it did not meet the goal, what we did to improve it, and what we will do to ensure it does not occur again. In this example we used the major output targets that the organization recognized as being critical to operation and started tracking them on a regular basis. The team may select more detailed goals such as scrap rate, rework percent, first time yield, attendance, overtime, or any other goal that is being tracked within the organization. Typically, each department will have a diagram that tracks all of their assigned goals. Each department head will present those diagrams at the regular management review meetings and compare their goals against all other departments within the organization.

SUMMARY
SUMMARY OF PROJECT –

Problem solving is not just a tool that can be applied to the manufacturing and service industries; the tools described in the previous chapters can be used in almost any situation that needs to be improved. Every problem is actually defined as a problem because someone is not satisfied. The reason the person is not satisfied is because they expected something different than what they actually received. If a couple in a relationship is arguing they are obviously not satisfied with the output they are receiving from the other. Notice how councilors or therapists seem to ask more questions than they actually answer? The reason is because they are attempting to define the actual problem the person or persons are having (remember the 5 why approach in Analyze phase). They are following the same problem solving road map as described in this project. They are attempting to figure out what the target is compared to what the person is actually receiving. For example: A long time married couple is arguing over a money issue and both parties think the other is wrong. Both of them stand their ground to the point that they are on the brink of destruction. The longer the problem goes on, the colder the trail gets and the harder it will be to gather the actual data for analysis. The actual problem was the wife got upset because the husband had taken money out of the joint savings account to pay for his portion of a scheduled fishing trip he had planned several months ago. To actually solve this problem we need to know where the actual results compare to the expected target goal that was established. The wife maintains that the money in the savings account was for a family vacation and the husband maintains that the money in the savings account was for both the family vacation and his fishing trip. The problem is now there is not enough money in the account to do both and the husband has already placed a huge deposit on his trip. Using the tools that have been gathered over this project, what do you see the problem is? Does it appear that both of them had different targets established for the joint savings account? Did both of them expect the same output? Obviously this is a very serious problem that could lead to many long term arguments between the couple for a very long time. When the joint savings account was set up neither the wife nor the husband clearly defined what the target goal was for the money. Both expected different outputs for the account and both were dissatisfied with the output. This is defined as a problem. Neither party is happy with the situation. Can this problem be solved? Of course it can be solved, but the problem has already been created and tension is now very high. It will be very difficult to create a win-win scenario until both parties are willing to cooperate and solve the problem systematically.

*We must first **define** the actual problem; quantify it: There is currently $1500 in the account.

*Next step is to take **measurement** of the problem: Husband needs $500 for his trip, family trip cost $1200.

*We now need to **analyze** the data: Husbands' trip actually cost $100 per day, total of 5 days. Family trip is a group rate of $1200 inclusive and can not be paid per day.

*We can now **improve** the condition: Husband agrees to cut his trip short by 2 days so that the additional money can be applied to the family trip.

*To finish this problem solving loop we need to ensure this does not reoccur by implementing **controls** to the process: The husband and the wife now plan all trips together with the money in the savings account. They agree to write down the goal statement on the deposit tracking book to remind them of what the money will be used for. They allocate the appropriate money for each trip that will be taken.

We have now come to the end of our project and the end of a very educational journey down the problem solving road. I hope every reader of this text now understands that problem solving is a group effort and should be taken on by everyone in the organization from shop floor worker to the CEO. It is very important to understand the three levels of the organization and how to approach each of them using the proper method to solve the problem, implement the solution, and control those improvements to prevent reoccurrence. Always remember that the path to solving the problem is just as important as solving the problem itself. It is not an answer to a question; it is a solution to the problem. We must fully understand the actual problem, determine exactly how bad the situation is, understand why it is the way it is, make improvements to eliminate the problem, and put in sound controls to ensure the problem goes away forever. We must document our path so that others can see the progress and success of our efforts.

Process mapping and using the YX diagram are very powerful tools that can be used by all levels of education throughout the organization. No matter what format selected, it is important to use those tools again and again. The more the team practices using the tools the more efficient problem solvers they will become. By using audits we are able to find those targets that are not performing as intended and see the gap from target to actual condition. We can implement a continuous improvement plan to address those issues found during auditing to ensure we sustain longevity within our organization and continue to compete within our market scope.

Hopefully this book has helped you gain a better understanding of how to solve problems on a more simple level using a systematic approach to do so. The more you use these tools the better you will become at using them. Problems, no matter how difficult or how simple still have one thing in common; they can all be solved given the adequate resources and attention. The best thing one can do to solve problems is to take one piece at a time and place it where it belongs. Remember that solving a problem is not simply answering the question; it is finding *The Solution to the Puzzle*.